TI-83/84 Plus and TI-89 Manual

Patricia Humphrey
Georgia Southern University

Intro Stats

Third Edition

Richard D. De Veaux
Williams College

Paul F. Velleman
Cornell University

David E. Bock

Boston San Francisco New York
London Toronto Sydney Tokyo Singapore Madrid
Mexico City Munich Paris Cape Town Hong Kong Montreal

Reproduced by Pearson Addison-Wesley from electronic files supplied by the author.

Publishing as Pearson Addison-Wesley, 75 Arlington Street, Boston, MA 02116.

ISBN-13: 978-0-321-49943-1
ISBN-10: 0-321-49943-3

4 5 6 BB 10 09

Forward

This calculator manual has been written to accompany the third edition of *Intro Stats* by Richard D. DeVeaux, Paul F. Velleman and David Bock. Its chapters have been organized to closely follow the order of topic presentation in the text. Material is primarily focused on the TI-83/84 series calculators, with additional instructions for the TI-89 series where its operation differs significantly from the others.

If, as the authors of the text assert, and I concur with their statement, statistics is a process involving three main steps: Think, Show, and Tell; then this book is primarily about the Show step. Calculators (and computers) are in essence tools that help with the mechanics of a process. The end results they give are numbers or plots. They cannot decide what procedure is desirable or valid for any given set of data. It is up to the individual to decide if the assumptions and conditions for the desired analysis are fulfilled; if not, the analysis is invalid and any conclusions drawn are bogus. The decision of what to do with a set of data is probably the biggest problem with this course; Statistics is not algorithmic like most mathematics courses a student has had up until now, but is largely driven by logic and interpretation. Similarly, the numbers given as results of a procedure are not the ending point. These must be interpreted in the context of the situation to draw appropriate conclusions.

In my examples, I have given full instructions in use of the calculators along with screen shots as appropriate to illustrate the inputs and outputs. I have also made appropriate conclusions as warranted. I have also included in each chapter, as appropriate, a What Can Go Wrong? section which illustrates common error messages and errors and tells how to correct these. These generally occur where an error might first be encountered.

In closing, let me assure you that using this manual *can* help you learn statistics – numerous of my students who have used earlier versions of this manual have said it was extremely helpful to them during the course. Not only that, but if my husband can learn from this, anyone can!

Lastly, I would like to say that it has been a pleasure working with the authors on this and the prior versions of the text. My thanks also to the Pearson Addison-Wesley people I have worked with: Deirdre Lynch, Joe Vetere, and Sara Oliver Gordus.

Contents

Chapter 1 – TI Calculator Basics

In this chapter we introduce our calculator companion to *Intro Stats* (3rd ed.) by giving an overview of Texas Instruments' graphing calculators: the TI-83, -83+, -84+, and -89. Read this chapter carefully in order to familiarize yourself with the keys and menus most utilized in this manual. You will also learn how to set the correct [MODE] on the calculators to ensure that you will obtain the same results as this companion does. You will learn other useful skills such as adjusting the screen contrast and checking the battery strength.

Aside from the above technical skills, you will learn some basic skills that are particularly useful throughout your study of *IntroStats*. Throughout this companion, we will present the uses of these calculators by illustrating their use on actual textbook examples or exercises. Since Chapter 1 is an introductory chapter in the text, we will take the opportunity in Chapter 1 of this companion to introduce you to skills which you will find necessary throughout the other chapters. These skills include Home screen calculations and saving and editing lists of data in the STAT(istics) editor.

KEY DIFFERENCES BETWEEN THE 83/84 SERIES AND THE 89 SERIES

All calculators in the TI-83/84/89 series have built-in statistical capabilities. Although a few statistical functions are "native" on the TI-89, most of the topics covered in a normal Statistics course require downloading the Texas Instruments Statistics with List Editor application which is free. Download requires the TI-Connect cable. The software (infstats.89g) can be found in the program archive on the TI website at ftp://ftp.ti.com/pub/graph-ti/calc-apps/89/math/stat/. This manual assumes the statistics application has been loaded on the calculator. If you have the newer TI-89 Titanium edition, the statistics application comes pre-loaded, and the TI-Connect cable is included with the calculator.

The TI-83 and -84 series calculators are essentially keystroke-for-keystroke compatible; however, the 84 does have some additional capabilities (some additional statistical distributions and tests, for example) with the latest version of the operating system, version 2.41 which is also available for download at http://education.ti.com/educationportal/sites/US/productDetail/us_ti84p.html. The regular TI-83 does not have the ability to use APPS (applications) which are in some cases extensive programs. If you have one of these regular calculators, you will not be able to use the APPS included on the CD-Rom accompanying the text to load data sets, but will have to key them in yourself. (If you have the cable and TI-Connect software, these can be loaded from the .txt files included on the CD in the same manner as 89's – see page 12 for details). Regular TI-83 users will, however be able to use the programs on the CD for such applications as analysis of variance and multiple regression.

There are some major differences in calculator operation and menu systems which will, in some cases necessitate separate discussions of procedures for the TI-83/84 and TI-89 calculators. Some of these become apparent in the next section. Not only are there differences between the three series, but there is also a difference in operation between the TI-89 and the TI-89 Titanium edition. On the Titanium, all "functions" on the calculator are essentially applications – when a regular TI-89 is turned on, the user is on the "home screen" similar to that for the TI-83 and -84. When the Titanium edition is first turned on, one must scroll using the arrow keys to locate the desired application – we'll say more about this later.

KEYBOARD AND NOTATION

All TI keyboards have 5 columns and 10 rows of keys. This may seem like a lot, but the best way to familiarize yourself with the keyboard is to actually work with the calculator and learn out of necessity. The keyboard layout is identical on the 83+ and 84+, and differs from the 83 by the substitution of the [APPS] key for the [MATRX] key. The layout of the 89 (and 89 Titanium) keyboard is similar, but some functions have been relocated. You will find the following keys among the most useful and thus they are found in prominent positions on the keyboard.

- The cursor control keys [◄], [►], [▲] and [▼] are located toward the upper right of your keyboard. These keys allow you to move the cursor on your screen in the direction which the arrow indicates.

- The [Y=] key is in the upper left of the 83 and 84 keyboards. It is utilized more in other types of mathematics courses (such as algebra) than in a statistics course; however you will use the [2nd] function above it quite often. This is the STAT PLOT menu on the 83/84 series. We will discuss [2nd] functions shortly. On TI-89 calculators, the Y= application is accessed by pressing [♦][F1]. The STAT PLOT menu on the TI-89 series is found inside the Statistics application.
- The [ON] key is in the bottom left of the keyboard. Its function is self-explanatory. To turn the calculator off, press [2nd][ON].
- The [ENTER] key is in the bottom right of the keyboard. You will usually need to press this key in order to have the calculator actually do what you have instructed it to do with your preceding keystrokes.
- The [GRAPH] key is in the upper right of the keyboard. On the TI-89, GRAPH is [♦][F3].

As mentioned briefly above, most keys on the keyboard have more than one function. The primary function is marked on the key itself and the alternative functions are marked in color above the key. The color depends on the calculator model. Below you will be instructed on how to engage the functions which appear in color.

The [2nd] Key

The color of this key varies with calculator model. On 83's and 89's this is a yellow key near the top left. On 84's and the 89 Titanium, the key is blue but is also at the top left. If you wish to engage a function which appears in the corresponding color above a key, you must first press the [2nd] key. You will know the second key is engaged when the cursor on your screen changes to a blinking ▮. As an example, on a TI-83 or -84 if you wish to call the STAT PLOTS menu which is in color above the [Y=] key, you will press [2nd] [Y=].

The [ALPHA] Key

You will also see characters appearing in a second color above keys which are mostly letters of the alphabet. This is because there are some situations in which you will wish to name variables or lists and in doing so you will need to type the letters or names. If you wish to type a letter on the screen you must first press the [ALPHA] key. The color of this key depends on the model of calculator: on the TI-83 it is blue; on the 84, green; on the 89, purple; and on the 89 Titanium, white. On 83's and 84's it is directly under the [2nd] key; it is one place to the right of that on both 89's. You will know the [ALPHA] key has been engaged when the cursor on the screen turns into a blinking ▮. After pressing the [ALPHA] key you should press the key above which your letter appears. As an example if you wish to type the letter E on an 83 or 84, press [ALPHA] [SIN] (because E is above [SIN]. To get the same letter E on an 89, press [ALPHA][÷].

Note: If you have a sequence of letters to type, you will want to press [2nd] [ALPHA]. This will engage the colored function above the [ALPHA] key which is the A-LOCK function. It locks the calculator into the Alpha mode, so that you can repeatedly press keys and get the alpha character for each. Otherwise, you would have to press [ALPHA] before each letter. Press [ALPHA] again to release the calculator from the A-LOCK mode.

Some General Keyboard Patterns and Important Keys

1. The top row on 83's and 84's is for plotting and graphing. On 89's these functions are accessed by preceding the desired function with [♦].
2. The second row from the top has the important QUIT function ([2nd] [MODE] on 83's and 84's, [2nd][ESC] on 89's). On 83's and 84's it also contains the keys useful for editing ([DEL], [2nd] [DEL] (INS), [◄], [►], [▲] and [▼]). INS and DEL on 89's are both combination commands: INS is [2nd][←] and DEL is [♦][←].

3. The [MATH] key in the first column on 83's and 84's leads to a set of menus of mathematical functions. Several other mathematical functions (like [x^2]) have keys in the first column. On a TI-89, [2nd][5] leads to the Math menu.
4. The keys for arithmetic operations are in the last column ([÷] [×] [−] [+]).

   Note: On all input screens, the [÷] shows as /, and the [×] shows as *. On both 89 models, when the command is transferred to the display area the * is replaced with a · and division looks like a fraction.
5. The [STAT] key, on 83's and 84's will be basic to this course. Submenus from this key allow editing of lists, computation of statistics, and calculations for confidence intervals and statistical tests. On 89's with the statistics application, one starts the application using the key sequences [◆][APPS] and selecting the Statistics application. On the 89 Titanium, quit the current application ([2nd][ESC]) and locate the `Stats/ListEditor` application, and press [ENTER] to start the application. On 83's and 84's the second function of the [STAT] key is `LIST`. This key and its submenus allow one to access named lists and perform list operations and mathematics.
6. The [VARS] key on 83's and 84's allows one to access named variables. On TI-89's this is [2nd][−] which is named [VAR-LINK]; it is used for both lists and variables.
7. [2nd][VARS] calls the distributions (`Distr`) menu. This is used for many probability calculations. To get this menu on a TI-89, press [F5] from within the `Stats/ListEditor` application.
8. The [,] key is located in the sixth row directly above the [7] key on 83's and 84's, while on 89's it is above the [9] key. It is used quite often for grouping and separating parameters of commands.
9. The [STO▸] key is used for storing values. It is located near the bottom left of the keyboard directly above the [ON] key on all the calculators. It appears as a → on the display screen.
10. The [(-)] key on the bottom row (to the left of [ENTER]) is the key used to denote negative numbers. It differs from the subtraction key [−].

Note: The [(-)] shows as ⁻ on the screen, smaller and higher than the subtraction sign.

SETTING THE CORRECT MODE

If your answers do not show as many decimal places as the ones shown in this companion or if you have difficulty matching any other output, check your `MODE` settings. Below we instruct on setting the best `MODE` settings for our work. These are the ones we have used throughout this companion.

On an 83 or 84, Press the [MODE] key (second row, second column). You should see a screen like the one on the right. If your calculator has been used previously by you or someone else the highlighted choices may differ. If your screen does have different highlighted choices use the [▲] and [▼] keys to go to each row with a different choice and press [ENTER] when the blinking cursor is on the first choice in each row. This will highlight and select that choice. Continue until your screen looks exactly like the screen to the right. Press [2nd] [MODE] (`QUIT`) to return to the Home Screen.

On TI-89's, the default mode is to give "exact" answers. For statistical calculations, you will want to change the mode to give decimal approximations. To set this option, press [MODE]. Press [F2] to proceed to the second page of settings, then arrow to Exact/Approx and use the right and down arrows to change the setting to `3:Approximate`. Press [ENTER] to complete the set-up. The sequence of screens is shown below.

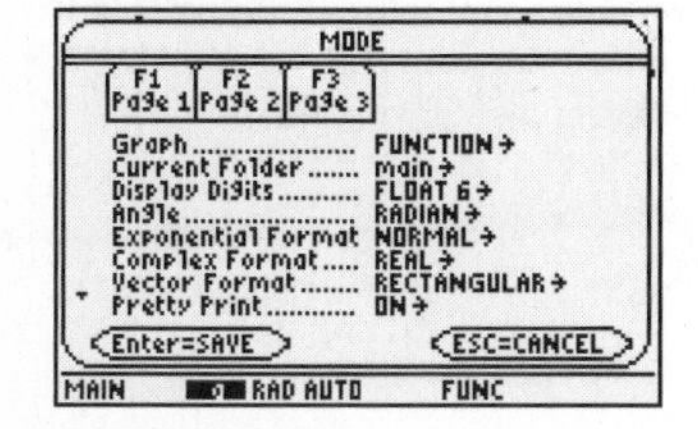

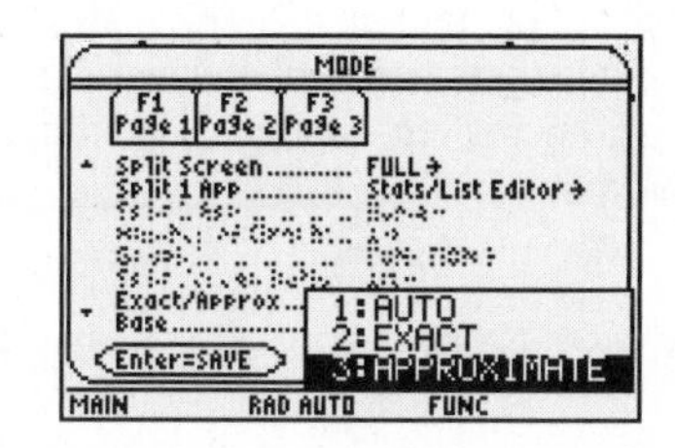

SCREEN CONTRAST ADJUSTMENT AND BATTERY CHECK

To adjust the screen contrast, follow these steps:

To increase the contrast on a TI-83 or -84, press and release the [2nd] key and hold down the [▲] key. You will see the contrast increasing. There will be a number in the upper-left corner of the screen which increases from 0 (lightest) to 9 (darkest). On a TI-89, press [◆][+].

To decrease the contrast, press and release the [2nd] key and hold down the [▼] key. You will see the contrast decreasing. The number in the upper-left corner of the screen will decrease as you hold. The lightest setting may appear as a blank screen. If this occurs, simply follow the instructions for increasing the contrast, and your display will reappear. On a TI-89, press [◆][−].

When the batteries are low, the display begins to dim (especially during calculations) and you must adjust to a higher contrast setting than you normally use. If you have to set the contrast setting to 9, you will soon need to replace the four AAA batteries. With newer versions of the operating system, your calculator will display a low-battery message to warn you when it is time to change the batteries. After you change batteries, you will need to readjust your contrast as explained above.

Note: It is important to turn off your calculator and change the batteries as soon as you see the "low battery" message in order to avoid loss of your data or corruption of calculator memory. Change batteries as quickly as possible. Failure to do so may result in the calculator resetting memory to factory defaults (losing any data or options which have been set).

A SPECIAL WORD ABOUT THE TI-89 TITANIUM

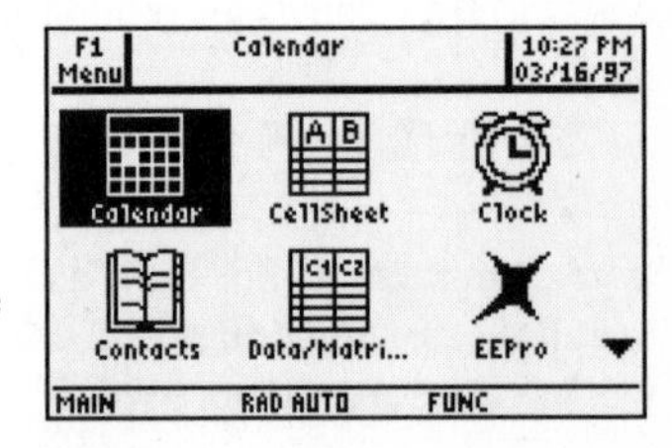

On the TI-89 Titanium, most important functions which on other calculators are accessed by keystrokes, are applications (Apps). When the calculator is first turned on, you will be presented with a graphical menu of these applications, as at right. Paging through the screen to find the one you want can be tiresome and time consuming. There is a way to customize this screen so that you only see those applications you want to see.

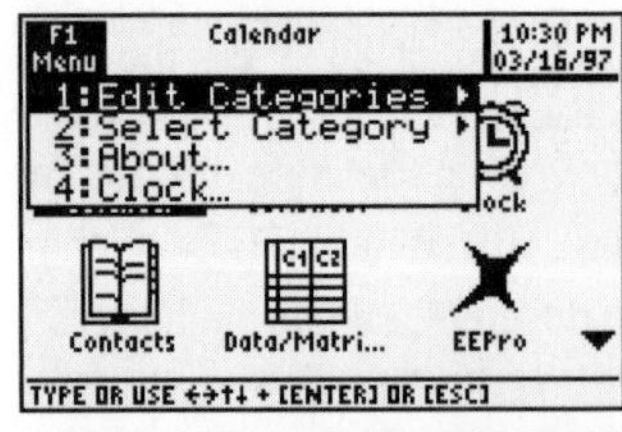

On the screen above, press [F1]. Press the right arrow key to expand menu selection `1:Edit Categories`. You will be presented with a list of possible categories. Press [3] to select option `3:Math`.

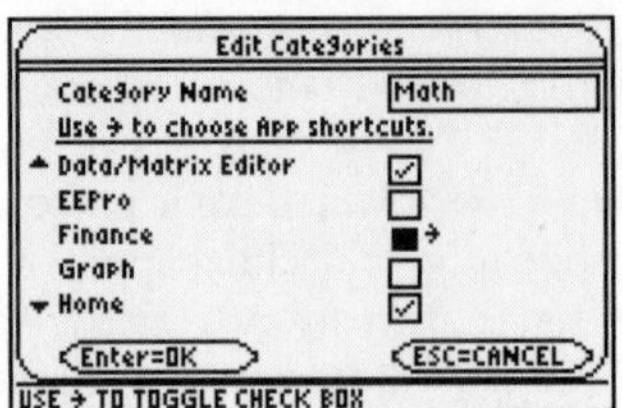

On this screen, use the down arrow to page through the list of applications. When you find one you want to be displayed, press the right arrow key to place a checkmark in the box. The screen at right shows that the `Data/Matrix Editor` and the `Home` screen have been selected. For this statistics course, you will want these applications, along with the `Stats/List Editor` and `Y=` applications. Press [ENTER] when you have finished making your selections.

On this calculator, pressing [2nd][ESC] (`Quit`) will return you to the applications selection screen. There are two useful shortcuts between applications. The first is pressing the [HOME] key, which takes you directly to the Home screen. The other useful shortcut is pressing [2nd][APPS] which allows you to toggle between two applications.

HOME SCREEN CALCULATIONS

The following examples illustrate some techniques which will be useful in performing home screen computations. These examples also point out the importance of correctly using parentheses in calculations.

Example 1: $\dfrac{98.20-98.60}{0.62}$

```
98.20-97..60
```

We will calculate the value in two ways. In doing so, we will intentionally make a mistake to show you how to correct errors using the [DEL] key. We will also discuss the `Ans` and Last `Entry` features.

Type 98.20 – 97..60 (an intentional mistake).

To correct the mistake, use the [◄] cursor key to move backward until your cursor is blinking on one of the double decimal points. Press [DEL] (on an 89, [♦][←] or position the cursor to the right of the character to be deleted and press [←]) and the duplicate decimal point will be deleted. Now press [◄] until the cursor is blinking on the 7. Type an 8, and it will replace the incorrect 7. On an 89, move the cursor to the right of the error, press [←] and then type the correct 8. Press [ENTER] for the numerator difference of ⁻4 as shown in the top of the screen below.

```
98.20-98.60
                 -.4
Ans/.62
       -.6451612903
```

Press [÷]. (Note that "`Ans/`" appears on the screen). Type .62 and press [ENTER] for the result of ⁻.645.
Note: `Ans` represents the last result of a calculation which was displayed alone and right-justified on the Home screen. Pressing [÷] without first typing a value called for something to be divided, so `Ans` was supplied.

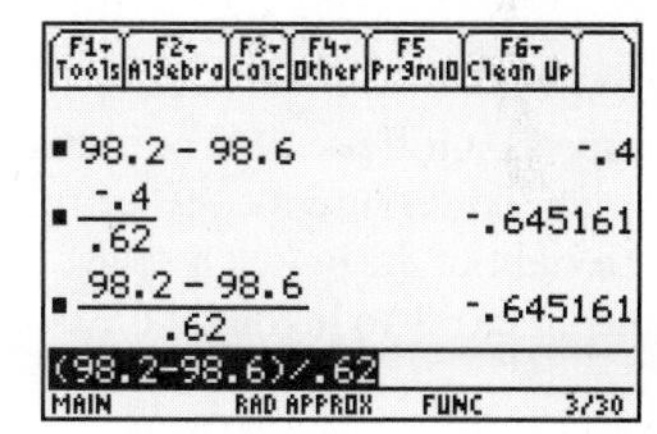

To do the calculation in one step, press [2nd] [ENTER]. This calls the "last entry" to the screen. (in this case Ans/0.62). Press [2nd] [ENTER] again to get back to 98.20-98.60. Press the [▲] key to move to the front of the line. On an 89, press [2nd][◄].

On an 83 or 84, you will need to press [2nd] [DEL] (for `INS`); 89's are always in insert mode. You will see a blinking underline cursor. Type [(] to insert a left parenthesis before the first 9. Press [▼] ([2nd][►] on an 89) to jump to the end of the line. Type [)] [÷] 0.62 to see the result. Press [ENTER] for the same result as before.

Example 2 $\sqrt{\dfrac{(5-7)^2+(12-7)^2+(4-7)^2}{3-1}}$

In this example, we will use the **ANS** function, illustrate syntax errors and show how to store quantities using variable names. Type (5-7) ^{2}+(12-7)^2+(4-7) ^{2} as in the screen. Press [ENTER] for the value 38. (Use the [x^2] key for the exponent 2.)

```
(5-7)²+(12-7)²+(
4-7)²
                  38
√(Ans/(3-1)
         4.358898944
```

Press [2nd] [x^2] [2nd] [(-)] [÷] and then type (3-1). Press [ENTER] for the desired results at the bottom of screen (8). Note that the [2nd] [x^2] sequence is the $\sqrt{\ }$ function on an 84 or 84; on an 89 it is [2nd][×]. The [2nd] [(-)] sequence calls the last answer, Ans back to the screen.

In the screen at right, we attempted to do the whole exercise in one step.

Pressing [ENTER] brings this message because we have made an error. Press 2 to "goto" the error.

We get this screen which has a blinking cursor on the last parenthesis. This means we have an extra right parenthesis which has no matching left parenthesis.

This screen shows the result when we go back and insert the missing left parenthesis into the calculation. We get the same result as before.

The screen at right shows the same calculation done on an 89 calculator. One important difference here is that the 89 does not have the [x^2] key. To exponentiate to any power, use the [^] key followed by the desired power. Also notice the [▶] at the right of the output display. This is a cue that there is more to be seen. Press the up arrow to highlight the output display, then press to right arrow to scroll to the end.

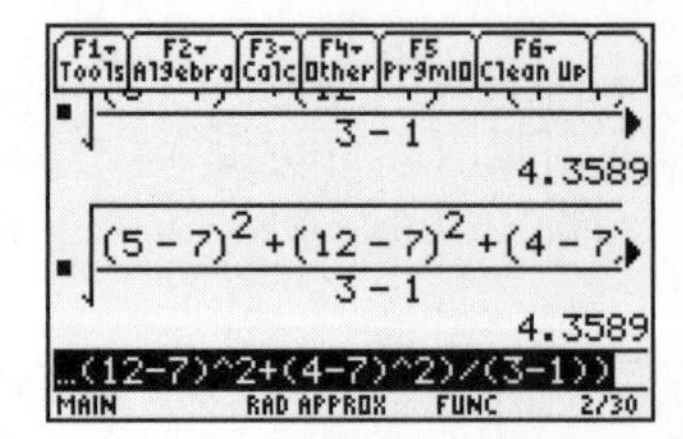

SHARING DATA

Sharing data between calculators (TI-83/84)

Data and programs may be shared between calculators using the communications cable which is supplied. The TI-83 and 84 series can share any TI-83/84 information with the exception of flash applications and their associated variables.

On the TI-83/83+, the I/O port is at the base of the calculator. On the TI-84/84+, you can use either the USB port or the I/O port on the top to link to another 84 series calculator. To link to an 83 series, you must use the I/O port. Connect the appropriate cable to the ports. On both calculators, press [2nd][X,T,Θ,n] to activate the LINK menu.

On the receiving calculator, press [▶] to highlight RECEIVE, then press

[ENTER]. The calculator will display the message "Waiting..."
The rolling cursor on the upper right indicates the calculator is working.

On the sending calculator, use the arrow keys to select the type of information to send. For example sending lists, either arrow to 4:List and press [ENTER] or press [4]. The screen at right will be shown. To select items to send, move the cursor to the item, press [ENTER] to select it. After selecting all items to send, press [▶] to highlight TRANSMIT, then press [ENTER].

Sorry, TI-83's and -84's cannot communicate with TI-89's.

Sharing Data between calculators (TI-89 Series)

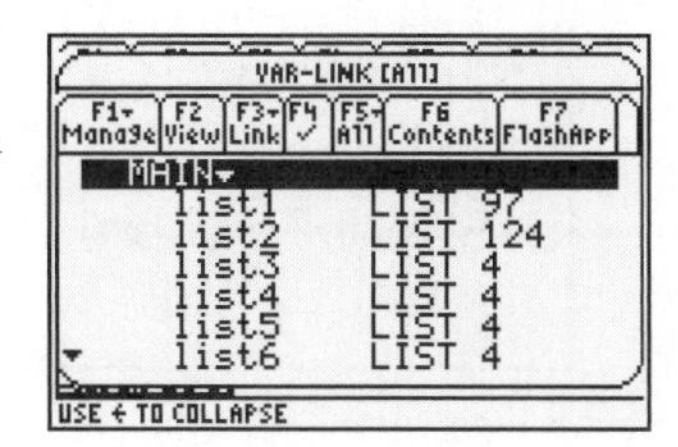

Data and programs may be shared between calculators using the communications cable which is supplied. The TI-89 can only communicate with the other TI-89's and TI-92s.

Connect the supplied cable to the port at the base of each calculator.
On both calculators press [2nd][-] to activate the VAR-LINK menu.

On the receiving calculator, press [F3] to select Link, ⊙ to highlight Receive, then press [ENTER]. The screen reverts to main VAR-LINK Menu with a "Waiting to receive" message at the bottom.

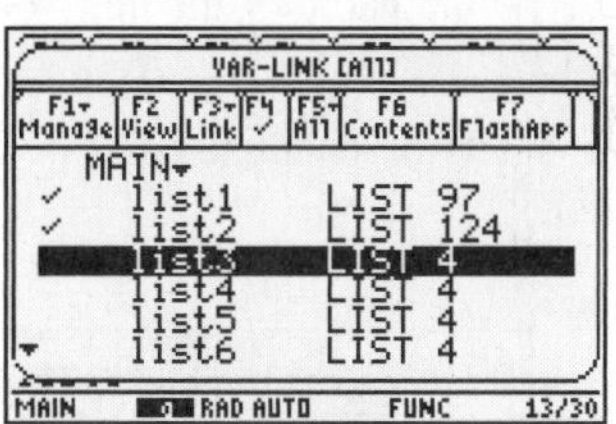

On the sending calculator, use the arrow keys to select the item to send, then press [F4] to "check" the item. The example at right will send list1 and list2.

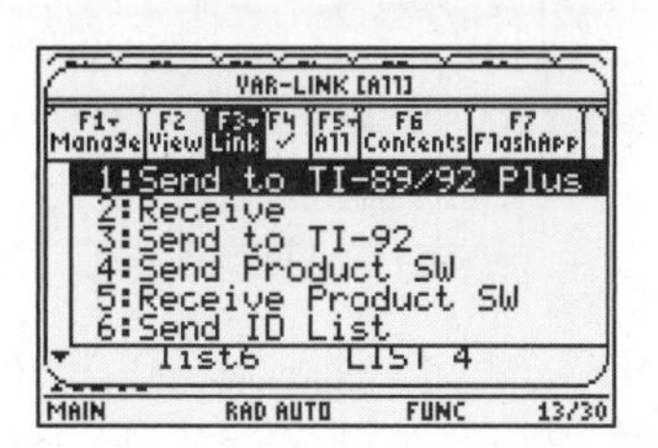

Now press [F3] to select Link, then [ENTER] to select menu choice 1:Send to TI-89/92Plus which is highlighted by default.

An analogous procedure can be used to send applications between calculators. Applications (such as the Statistics with List Editor) are selected from the [F7]FlashApp menu (press [2nd][F2] for [F7].)

Sharing data between the calculator and a computer

Data lists, screen shots and programs may be shared between the calculator and either Microsoft Windows or Macintosh computers using a special cable and either the TI-Connect or TI-Graph Link software. Cables for the TI-83 are available for either serial or USB computer ports and can be found through many outlets such as OfficeMax, and Amazon.com. The software can be downloaded free through the Texas Instruments website at education.ti.com. The needed USB cable and computer software are included with the TI-84 and -89 models.

WORKING WITH LISTS

The basic building blocks of any statistical analysis are lists of data. Before doing any statistics plot or analysis the data must be entered into the calculator. The calculator has six lists available in the statistics editor; these are `L1` through `L6` (`list1` through `list6` on an 89). These are accessed and referred to pressing [2nd]*n* on an 83/84 where *n* is the number of the list. On an 89, all lists are accessed through `VAR-LINK` which is [2nd][-]. Other lists can be added if desired. The number of lists is only limited by the memory size.

The Statistics Editor

On a TI-83/84, press [STAT], `1:Edit...` will be highlighted. Press [ENTER] to select this function.

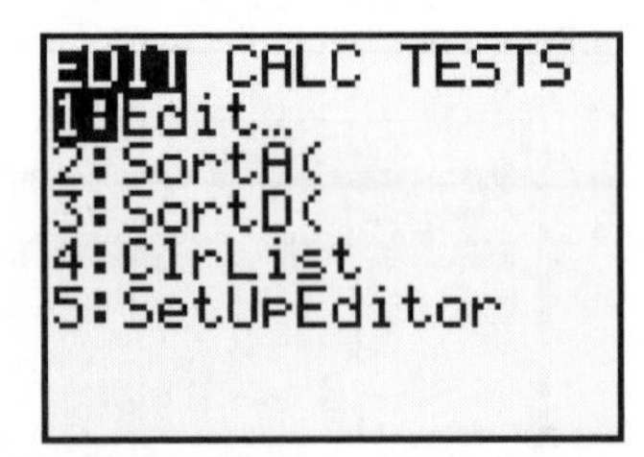

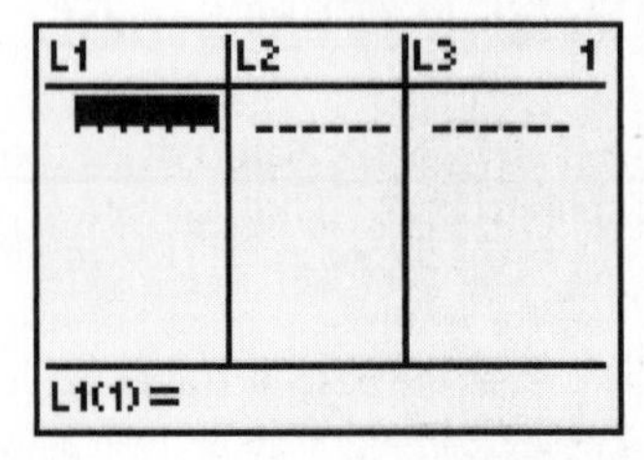

To enter data, simply use the right or left arrows to select a list, then type the entries in the list, following each value with [ENTER]. Note that it's not necessary to type any trailing zeros. They won't even be seen unless decimal places (found with the [MODE] key) have been set to some fixed number.

On a TI-89, press [♦][APPS] followed by the selection of the `Statistics/List Editor` Flash Application followed by [ENTER]. On the Titanium edition, select the `Statistics/List Editor` from the main application menu screen. If this is the first time the editor has been accessed since the calculator has been turned on, you will be prompted for a data folder as in the middle screen. The default folder is `main`. Press [ENTER] to select `main` as the current folder, or press ⊙ to allow a new folder to be created. To enter a new folder name, arrow to the entry block and type the name of the new folder. Your lists will be stored in the new folder and it will be set to be the default. To change folders, press ⊙ to select a folder. Then press [ENTER] to proceed to the list editor. If the editor has been used since the calculator has been turned on, pressing [ENTER] to select the application will automatically open the editor.

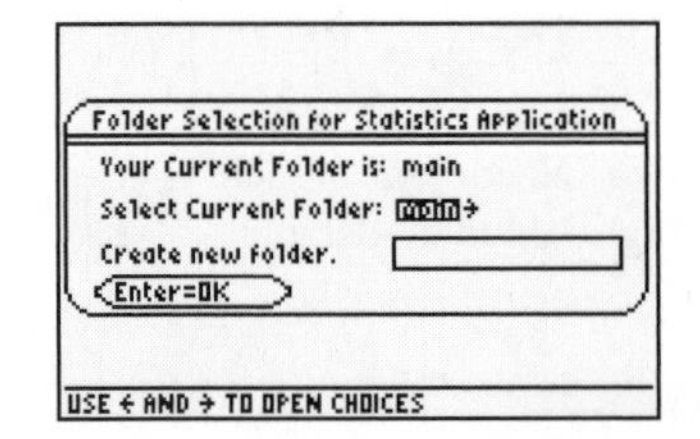

list1 list2 list3 list4
list2[1]=
MAIN RAD AUTO FUNC 2/7

One word of advice: Most lists of data in texts are entered across the page in order to save space. Don't think that just because there are four (or more!) columns of data they belong in four (or more) lists. Data which belongs to a single variable always belongs in a single list.

Entering Data into the STAT Editor

L1	L2	L3 1
1	31	------
3	1	
4	2	
45	11	
5	27	
	52	
	47	

L1(6)=

With the cursor at the first row of `L1`, type 1 and press [ENTER]. The cursor moves down one row. Type 3 followed by [ENTER] and the 3 will be pasted into the second row of the list. Continue with 4, 45, and 5 as seen at right.

Correcting Mistakes with DEL and INS

In the screen above, we can delete the 45 by using the ▲ key until it is highlighted and then pressing DEL (◆ ← on an 89).

To insert a 2 above the 3 move the cursor to the 3 then press 2nd DEL (2nd ← on an 89) (to choose INS or insert mode). Note a 0 was inserted where you wanted the 2 to go. Just type over the place-holding 0 with the value you want.

Clearing Lists without Leaving the STAT Editor

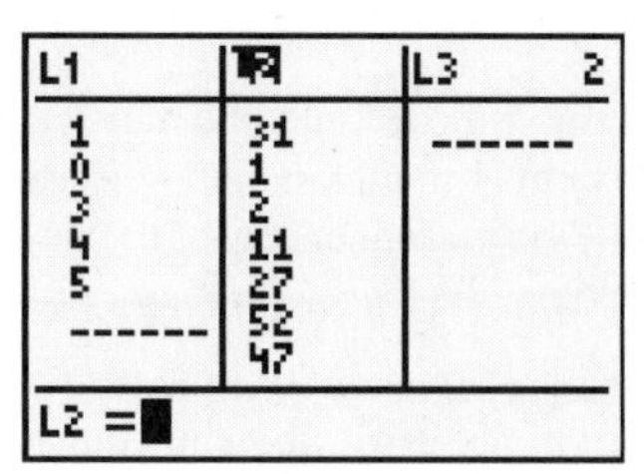

Suppose you wish to clear a list, say L2, while you are still in the STAT Editor. You should use the cursor to highlight the name of the list at the top. With the name highlighted, press CLEAR and you will see this. Press ENTER and the contents of the list will be cleared. *Make sure not to press* DEL or the list will be deleted entirely and you will have to use SetUpEditor as described below to retrieve it.

Deleting a List from the STAT Editor:

If you wish to delete a list from your STAT Editor, simply highlight the list name and press DEL. The name and the data are gone from the Editor but not from the memory. To recover a list inadvertently deleted, use SetUpEditor as described below.

SetUpEditor

Setting up the editor will remove unwanted lists from view. It also will recover lists which have inadvertently been deleted. On an 83 or 84, if you want the STAT Editor to be restored to its original condition (with lists L1 to L6 only), press STAT 5 ENTER. Often students find this necessary because they have inadvertently deleted one of the original lists.

On an 89, in the Statistics Editor, press F1 (Tools), then select option 3:Setup Editor. You will see the screen at right. Leave the box empty and press ENTER to return to the six default lists.

Using SetUp Editor to Name a List:

On a TI-83 or -84 home screen, press STAT 5 to call SetUpEditor. Now type IDS, STU, L1. You will have to be careful to keep pressing the ALPHA key before each character or release the alpha lock (2nd ALPHA) to type the commas. Then press ENTER. Then press STAT 1 to view your lists.

In the screen at right, you will see an old list STU has been placed back in full view. Also there is a new list IDS ready and waiting for some data to be entered. Try pressing the right arrow to see more lists – there aren't any. This command set the editor with just the three lists specified.

To do this on an 89, follow the steps outlined above to generally set up the editor, and type in the list names just as one does for the 83 or 84.

Generating a Sequence of Numbers in a List

From time to time one may want to enter a list of sequenced values (years for example in making a time-series plot). It is certainly possible (but tedious) to enter the entire sequence just as one would enter normal data. There is an easier option, however. Use cursor control keys to highlight L1 in the top line. Press [2nd] [STAT] [▸] [5]. You are choosing the LIST menu and then choosing the OPS submenu. From the OPS submenu you choose option 5 which is seq(.

This has been pasted onto the bottom line of the screen. Type in the rest so that you have seq(X,X,1,28. Press [ENTER] and the sequence of integers from 1 to 28 will be pasted into L1 as in my screen. Find the X on the [X,T,Θ,n] key on an 83 or 84, on 89's it has its own key.

On an 89, the procedure is analogous, but access the LIST OPS menu by pressing [F3][2], then select option 5.

Note: To quickly check the values on a multi-screen list you can press the green [ALPHA] key followed by either the [▴] or [▾] key. This will allow you to jump up or down from one page (screen) to another. The green arrows on the keyboard near the [▴] and [▾] keys are there to remind you of this capability. On a TI-89, instead of the [ALPHA] key, press [♦].

Sorting Lists (TI-83/84)

Lists may be sorted in either ascending (smallest to largest value) or descending order. The resulting list will replace the original list. On the main [STAT] menu select either 2:SortA(or 3:SortD(. The command will be transferred to the home screen. Enter the name of the list to be sorted ([2nd]n where n is the number of the list). Execute the command by pressing [ENTER]. The example below sorts list L1.

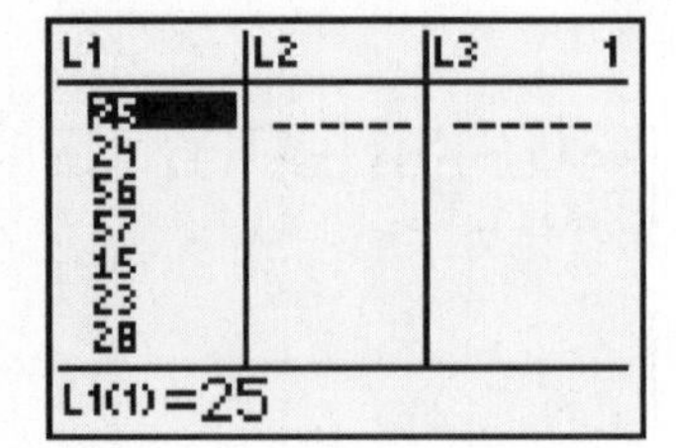

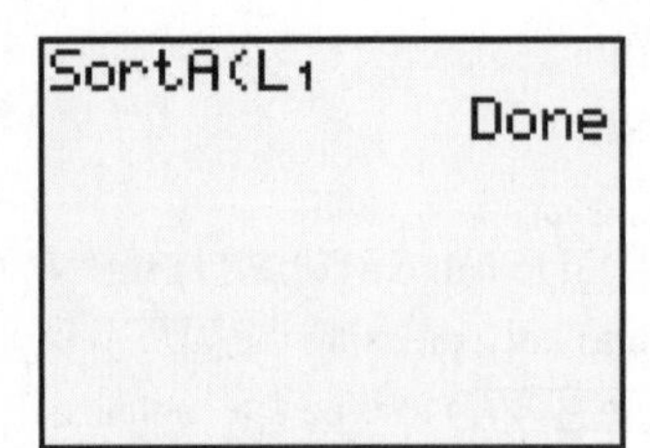

Sorting Lists (TI-89)

While in the List Editor, press [F3] for the `List` menu, press ⊖ followed by [ENTER] or [2] to select `List Ops`, then press [ENTER] to Select `1:Sort List`, [2nd][-] (`Var-Link`) and use the arrows to select the list to be sorted. Press the right arrow if necessary to change the sort order (use ▶ to activate the menu choices). Press [ENTER] to carry out the command.

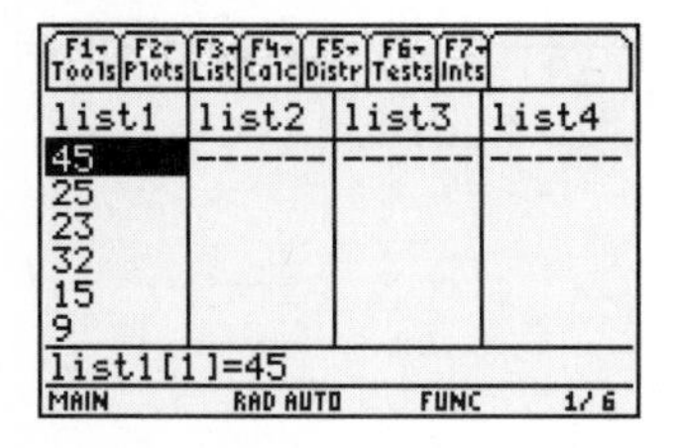

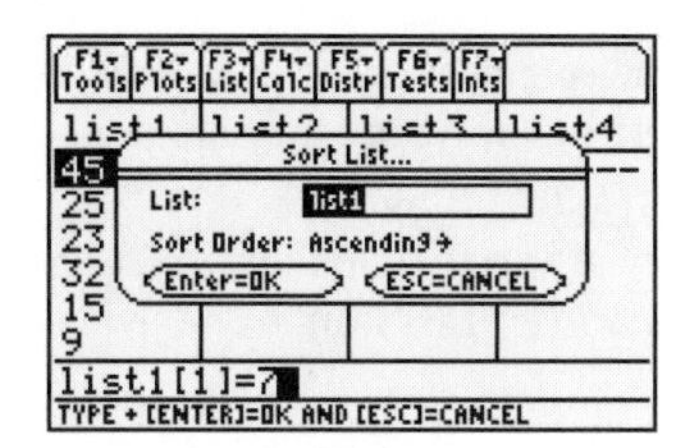

USING THE SUPPLIED CD-ROM DATA SETS

TI-83/84 Procedure

The CD-Rom supplied with the text includes applications which will load pre-entered lists of data for use with either the TI-83+ or TI-84 series calculators. (Regular TI-83 calculators do not have the application feature). Data sets are also available on the companion web site wps.aw.com/aw_deveaus_stats_series. To use the applications, load the CD on the computer; it should automatically start and present an initial *ActivStats* screen. There is a button on the lower left of the screen which is labeled Datasets. Click on the Datasets button, then click on the TI-applications folder. There is an application for each chapter of the text. With the calculator connected to the computer (with the graph-link or direct USB cable for TI-84's), right click on the desired chapter's application, and select "Send to TI device" from the pull-down menu. An interim screen will show the desired application. Click the "Send to Device" button at the bottom of the screen. If sending the app to a TI-84, nothing will display. If sending to a TI-83+, you will see a message that the calculator is receiving the application. When finished, the calculator screen will be blank and the transfer window on the computer will disappear.

To use the lists, start the application for the appropriate chapter. Press [APPS]. Locate the application for the chapter you want. The screen shown at right will begin the Chapter 4 application.

You will briefly see an introductory screen, then the one at right. Press [ENTER] to proceed. This is mainly to verify that you have the correct chapter's application selected. If you have selected the wrong chapter, press ▶ to highlight `Quit` and press [ENTER] to exit the application.

The next screen works similarly to the one before. To proceed to the exercise datasets press [ENTER], or press ▶ to highlight `Quit` and press [ENTER] to exit the app.

Now we can select the data sets to load into active memory. The lists are all named mnemonically and listed alphabetically to correspond with the subject of the problem. Select lists to load by moving the cursor to the list name and pressing [ENTER]. When finished selecting lists, press the right arrow key to highlight `Load` and press [ENTER]. The screen at right will load the lists associated with Acid Rain (problem 45), and Bird species (problem 40).

After moving the cursor to `Load`, you will see the screen at right. You have two options. Selecting `2:Load` will load the lists into memory, but they are not accessible except through the [LIST] menus ([2nd][STAT]). If you select `1:SetUpEditor` the first selection (`1:AddtoEditor`) will allow you to add these the selected lists at the end of the normal Statistics List Editor.

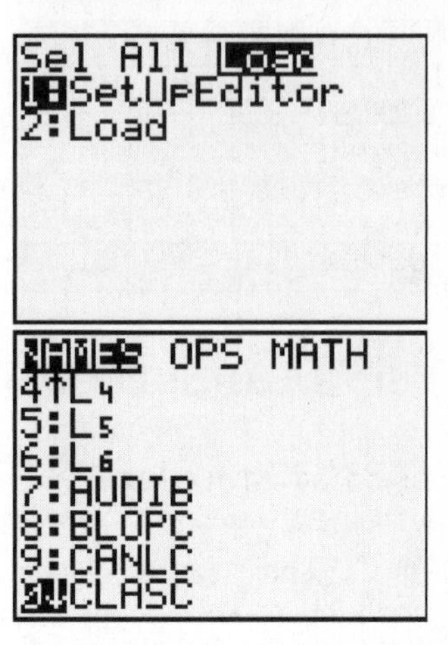

When using the lists for graphics or statistics, you will have to access the lists by name through the `LIST` menu ([2nd][STAT]). Any time one would normally give a default `L1` through `L6` list name by pressing [2nd]n, where n is the number of the list, press [2nd][STAT] for the list names screen, then arrow down to place the cursor at the appropriate list name and press [ENTER] to select it.

TI-89 Series Procedure

The CD-Rom supplied with the text includes data sets which can be downloaded into TI-89 calculators (they are also included on the book's website). If you have the connect cable and TI-connect software, this can save the time involved in entering the data manually. To use the data sets, load the CD on the computer; it should automatically start and present an initial *ActivStats* screen. There is a button on the lower left of the screen which is labeled Datasets. Click on the Datasets button, then click on the TI Files folder.

There are folders for each chapter of the text. Click on the appropriate chapter. The files are text (xxxxx.txt) files, which are named mnemonically to agree with the subject matter of problems. For example, in Chapter 4, acid rain is the subject of problem 45 and its data set is acidr.txt; the number bird species is the subject of problem 40 and its data set is birds.txt. Click on the appropriate data set and it will be opened (on a PC) by Notepad. Use the mouse and drag to highlight the list of data. Press CTRL-C (or Edit, Copy using the pull-down menu) to copy the data.

With TI-Connect, open the TI Data Editor. Click the blank page icon to open a new variable. Click to place the cursor at the top of the list (where there is a 0 placeholder), and press CTRL-V (or Edit, Paste from the pull-down menu) to paste the data into the list. Click in the beige area at the top of the list, and you will be prompted to give the list a name of up to six characters. Once the list is named, click Actions on the menu bar, then click Send Selected Items. The list will be sent to the TI-89 calculator and be given the name you have selected. Use the list just as you would any other.

MEMORY MANAGEMENT

Too many applications loaded or lists in active memory can overload the calculator. Just as a computer disk can be filled up, so can memory on the calculator. The TI calculators have two types of memory – RAM and archival. If you use the applications supplied on the CD (on a TI-83/84), these are loaded into archival memory. Active lists are in RAM.

TI-83/84 Procedure

To find out the current free memory status, press [2nd][+] (MEM) and select option 2:Mem Mgmt/Del.

The screen at right shows my calculator currently has 18,927 bytes (characters) of free RAM and 1303Kilobytes of free archival memory.

If you need to free some memory, decide the type. If you want to delete some lists, for example, select 4:List. Move the cursor to the lists you wish to delete and press [DEL] for each one. The can also be done for any applications you no longer need from previous chapters, but use choice 5:Archive to access the list of archived applications.

TI-89 Series Procedure

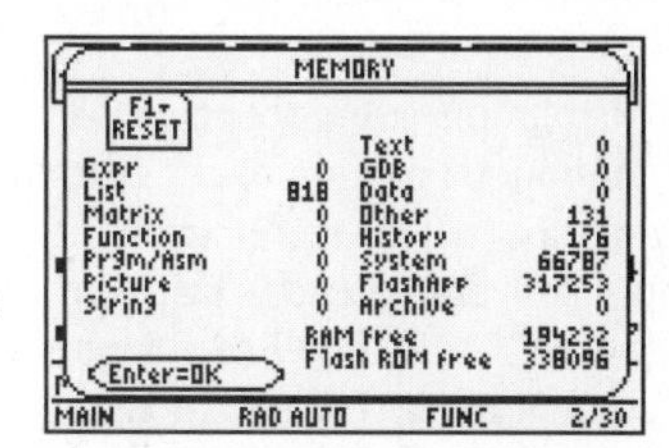

To find out the current free memory status, press [2nd][6]. The screen at right shows I currently have 194,232 free bytes of RAM and 338096 free bytes of Flash memory free. Pressing [F1] here will reset RAM, Flash, or all memory to either totally blank or factory default settings. I do not recommend either of these options under normal circumstances. Resetting Flash, for example would erase the Statistics Flash application, which would then need to be reloaded.

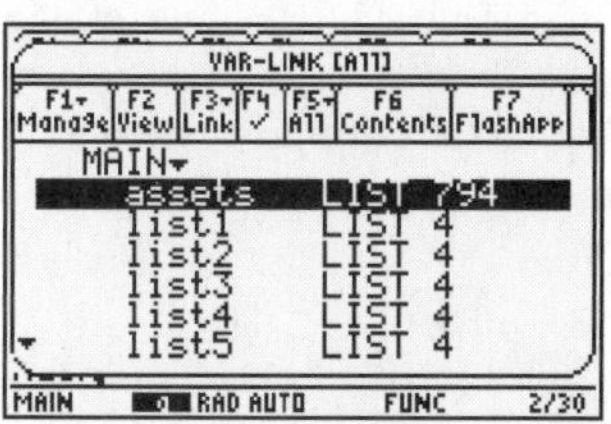

To delete lists which are no longer needed, press [2nd][-] (VAR-LINK). Move the cursor to highlight the list to be deleted, then press [F1]. Press [ENTER] to select option 1:Delete.

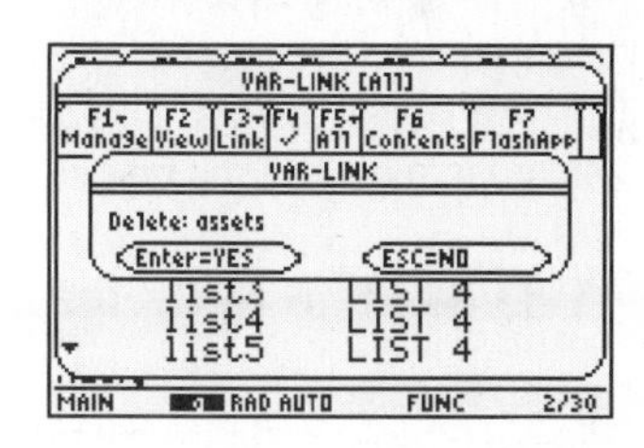

You will be prompted to verify that the selected item is to be deleted. Press [ENTER] to confirm the deletion, or [ESC] to cancel. You can continue this process to delete all unneeded lists.

WHAT CAN GO WRONG?

Why is my list missing?

By far the most common error, aside from typographical errors is improper deletion of lists. When lists seem to be "missing" the user has pressed [DEL] rather than [CLEAR] in attempting to erase a list. Believe it or not, the data and the list are still in memory. To reclaim the missing list press [STAT] and select choice 5:SetUpEditor followed by [ENTER] to execute the command. Upon return to the Editor, the missing list will be displayed.

Chapter 2 – Displaying and Summarizing Quantitative Data

The three primary rules of data analysis are:

1. Make a picture.
2. Make a picture.
3. Make a picture.

TI-calculators can help with this, although they cannot make bar graphs or pie charts for categorical data or dotplots and stem-and-leaf displays. All of these are fairly easily done by hand (at least for small data sets; for larger ones, use a full computer package).

This chapter will discuss histograms for numeric data. Other plots of quantitative data will be discussed later. Histograms are examined for shape (skewed right/left or symmetric), center, and spread. They also tell us whether or not a distribution is unimodal (one-humped) or multi-modal (many humps). We will also look at computing summary statistics for a set of data.

HISTOGRAMS

Histograms are connected barcharts. Since the data are presumed to represent particular observations on some continuous portion of the real number line and since order here matters, bars are always displayed connected to one another (unless there happens to be a gap in the values). A good histogram has equal bar widths, high and low ends not too dramatically different from the maximum and minimum values of the data, and intervals which "make sense." As we will see, they are useful in highlighting major features of the distribution of a single variable or for comparing two distributions (if done properly); they also have a capacity for "artiness" since their shape can change, depending on the choice of beginning values and bar widths.

In this first example, we will use the data on pulse rates of 24 women to create the histogram shown in figure 4-3 of the text. For convenience, the data are below.

88	56	80	60	76	72	68	80
64	80	84	64	68	72	80	76
72	76	84	76	72	68	68	64

The first step in making a histogram is to enter the data. The data have been entered into list `L1`; the first few values are seen in the accompanying figure.

TI-83/84 Steps to Create a Histogram

The next step is to define the plot. This is done by pressing [2nd][Y=] (`Stat Plot`). You will see the screen at right. Notice that there are three plots which can be displayed at any one time. For most purposes, there should be only one turned "on" at once. Notice here `Plot1` is `On` and Plots 2 and 3 show `Off`. Scrolling down the menu are options 4 and 5 that turn all plots off or on with a single command. Selecting either of these will transfer the command to the home screen. Executing it requires pressing [ENTER].

Press [ENTER] to select `Plot1`. The cursor should be blinking over the word `On`. If `On` is not already highlighted, press [ENTER] to move the highlight. Notice that there are six graphics types. Histograms are the third choice. Pressing [▼] will move the cursor to the first plot type. Use [▶] to move the cursor to the histogram figure. Press [ENTER] to move the highlight.

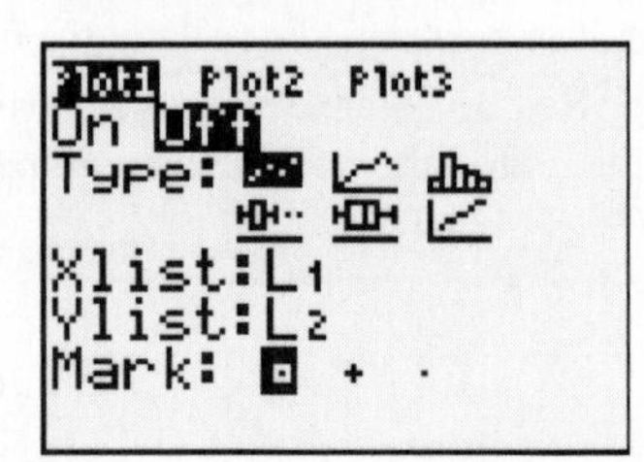

At this point, your screen probably looks like this. We're ready to display the graph, since our data was in list `L1` and each data value had frequency 1 (represented one observation). If you want to graph data in other lists, move the cursor to `Xlist:` and enter the list name ([2nd] n, where n is the number of the list). We'll talk more about frequencies later. Notice if you move the cursor to `Freq:` it will flash as A. If you need to change this back from something else to a 1 you will need to press [ALPHA] before typing the 1.

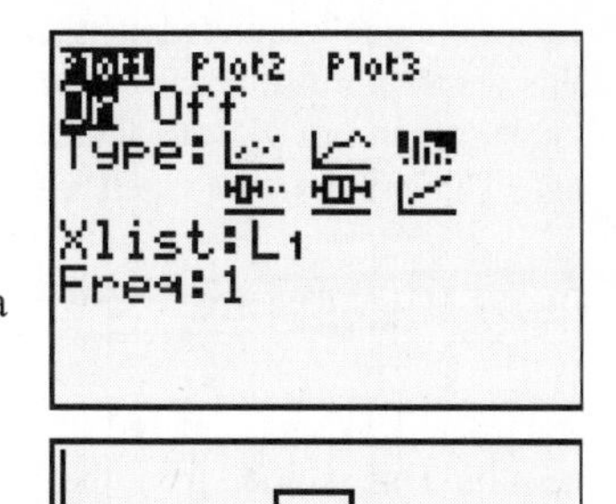

The easiest way to display a histogram (or any statistics plot) is to press [ZOOM][9] (`Zoom Stat`). The resulting graph is seen at right. Notice the *y*-axis penetrating one of the bars. The *x*-axis "floats" a little way up from the bottom of the screen. This is so that values as seen in the next picture do not interfere with the plot.

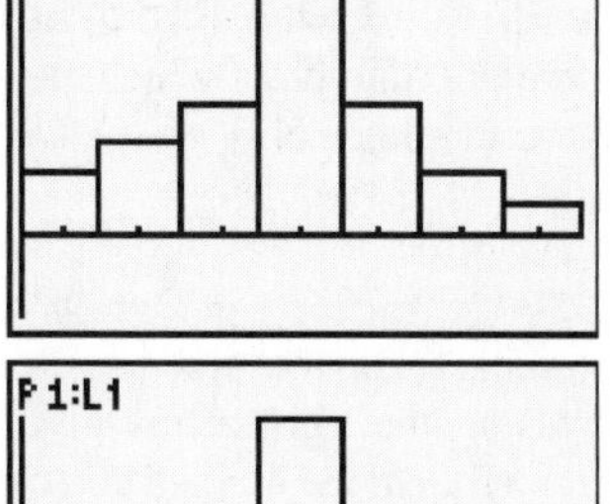

To see exactly what the graph shows, press [TRACE]. A blinking cursor will appear in the leftmost bar. At the bottom of the screen the minimum and maximum values for the bar, and number of observations in the bar are displayed. This first bar goes from 56 to 61.3333. There are two observations in this interval, indicated by the n=2 at the lower right.. Pressing the right arrow key ([▶]) will allow you to continue through the graph.

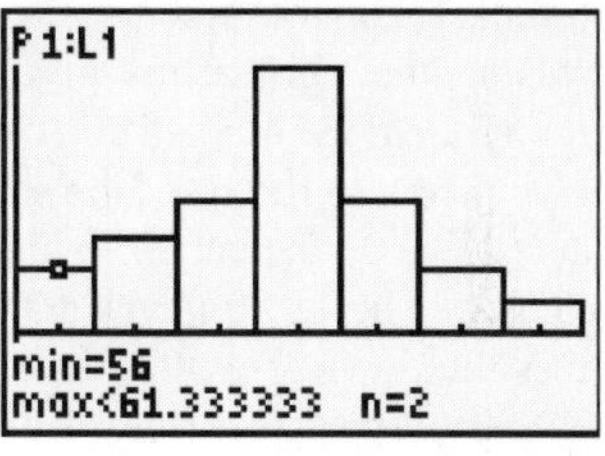

At this point, we can see the distribution of pulse rates appears to be unimodal and relatively symmetric (bars fall roughly equally from the center peak.) We see the center is around a 75 beats per minute with 8 observations in that bar.

There is a downfall to using simply [ZOOM][9] for histograms. Look at the first interval. It doesn't really make sense in a natural way. The bar width represents a difference between the low and high ends of each bar of 5.333333… which is unnatural. We'd like to fix this.

Manipulating Windows

To force particular minimums, maximums and scaling we will press [WINDOW]. This displays a screen like the one at right. Notice the `Xmin` was the smallest value shown on the plot and `Xscl` was the bar width. These are quantities we'd like to change. You generally won't have to change any of the *Y* variables here (unless a scaling change loses the top of a bar – then increase `Ymax`). Another reason to possibly change *Y* variables is to increase resolution.

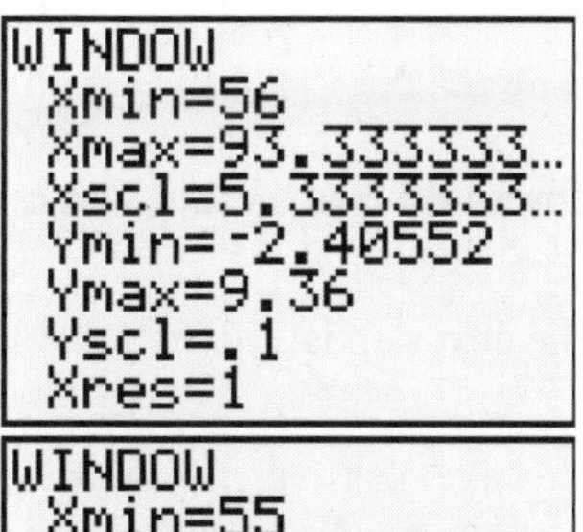

Change `Xmin` to 55 and `Xscl` to 5 (sounds pretty reasonable, and reproduces the scaling used in the text.)

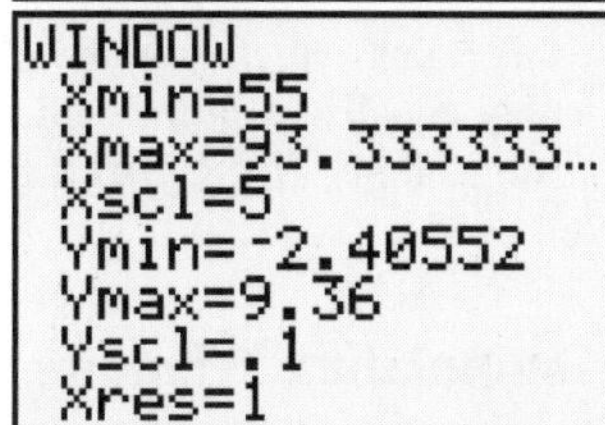

To display the new graph, press [GRAPH]. NEVER press [ZOOM][9] after changing a window. You'll just go back to the one you had before! This looks better, but changing the scaling left some room at the top of the graph. Let's change the window and lower the `Ymax` to 7.

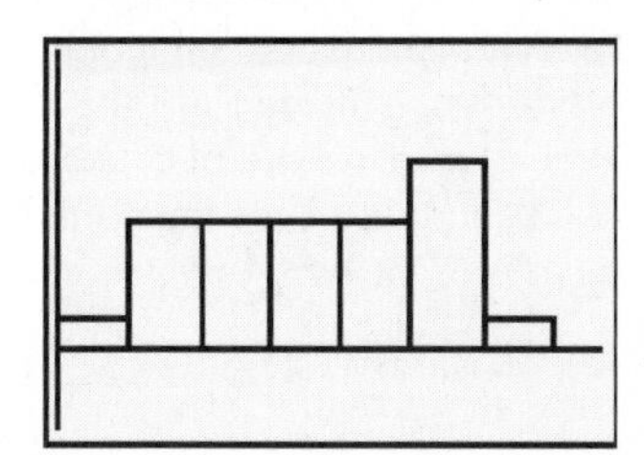

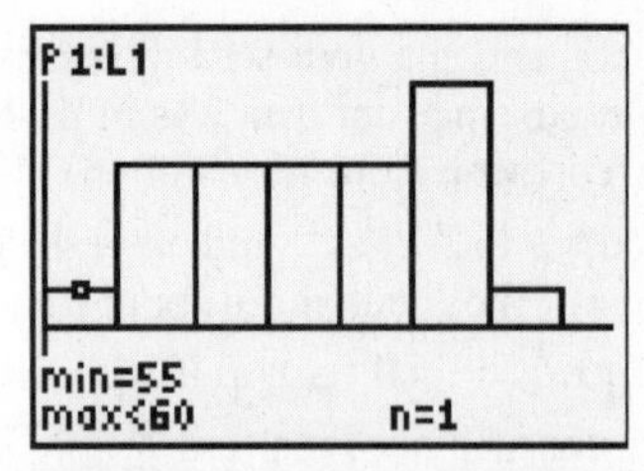

Here's the new graph (remember, press [GRAPH] after changing the window again). This graph points out one of the pitfalls of histograms as mentioned above – you notice that in this scaling, the central peak seen in the default histogram has been lost. This graph looks much more uniform (no real peak in the graph).

TI-89 Histograms

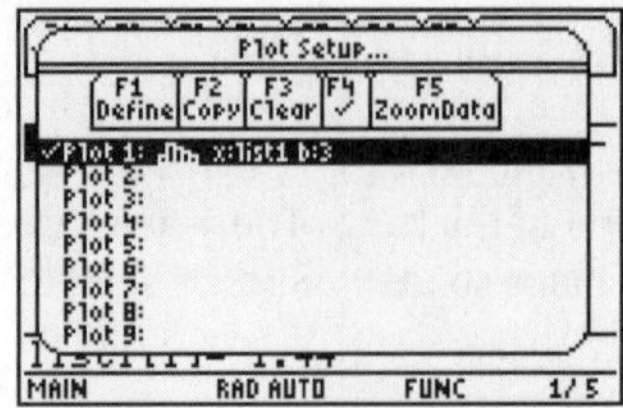

Once data has been entered in the Statistics/List Editor, the next step is to define the plot. This is done from the Statistics Editor by pressing [F2] (Plots) followed by [ENTER] to select Plot Setup. You will see the screen at right. Notice that there are nine plots which can be displayed at any one time. For most purposes, there should be only one active (checked) plot at once.

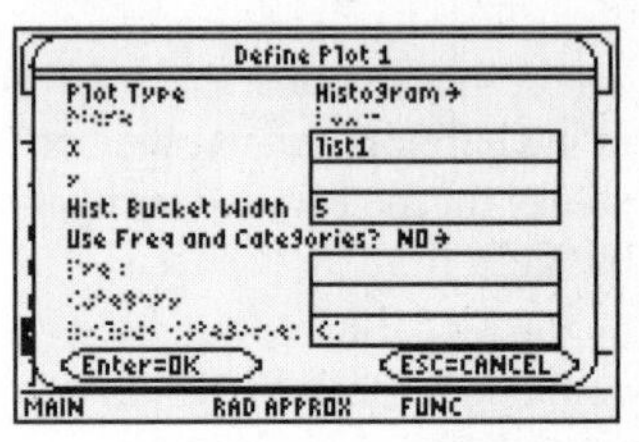

Press [F1] to select defining Plot1 since it was highlighted. The cursor should be blinking over the plot type. If the plot type is not already set to a histogram, pressing the right arrow gives a menu of five plot types. Move the cursor to highlight choice 4:Histogram and press [ENTER] to select it or press [4].Press the down arrow to the box labeled x. Press [2nd][−] (VAR-LINK) to get the list of list names. Move the cursor to highlight the one you wish to use, then press [ENTER] to select it. The TI-89 then wants the histogram bucket width which is the bar width. Press the down arrow to move to this box. You may need to use trial and error to get a good picture. Here, I have set the bar width to 5. Since we don't have a separate list of frequencies, Use Freq and Categories is set to NO. Press [ENTER] to complete the plot definition. You will be returned to the Plot Setup menu.

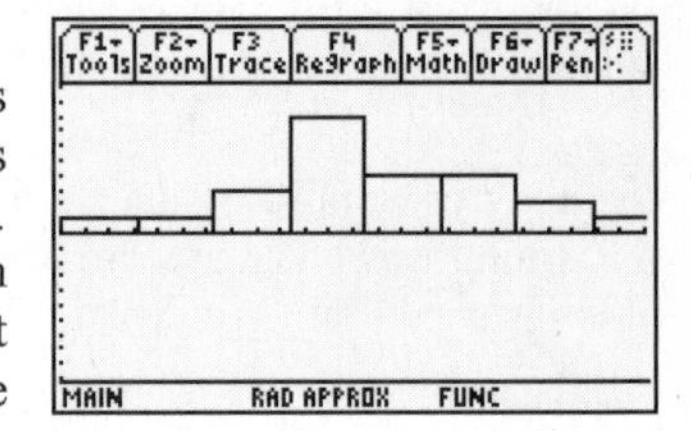

The easiest way to start displaying a histogram (or any statistics plot) is to press [F5] (Zoom Data). The resulting graph is seen at right. Notice the *y*-axis penetrating one of the bars. The *x*-axis "floats" up from the bottom of the screen. This is so that values seen in the next picture do not interfere with the plot. In many cases, the TI-89 will not show the full heights of the bars due to the tabs at the top lf the screen. It does not always get the windowing correct. We don't see the full heights of the bars. We'll change that later.

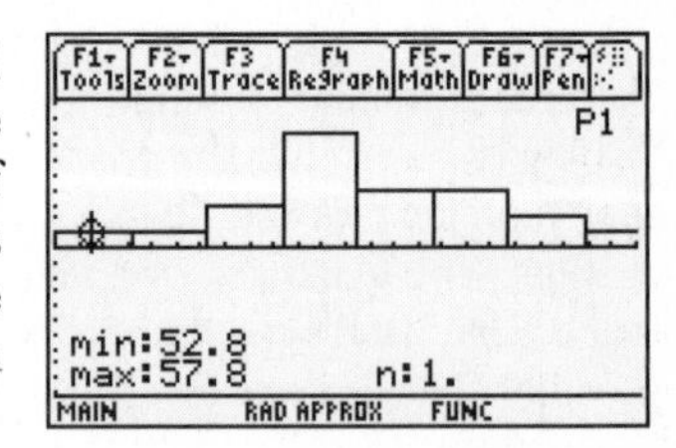

To see exactly what the graph shows, press [F3](Trace). A blinking cursor will show in the first bar at the left of the graph. At the bottom of the screen are displayed the minimum value and maximum value for the bar, and the number of observations in the bar. This bar goes from 52.8 to 57.8. There is one observations in this interval. Pressing the right arrow key (▶) will allow you to continue through the graph seeing the interval ranges and numbers of observations in each interval.

At this point, we can see the distribution of pulse rates appears to be unimodal and relatively symmetric (bars fall roughly equally from the center peak.) We see the center is around 73 beats per minute. There is another downfall to using simply Zoom Data for histograms. Look at the intervals. They really do not make sense in a natural way. Well need to fix this.

Manipulating Windows

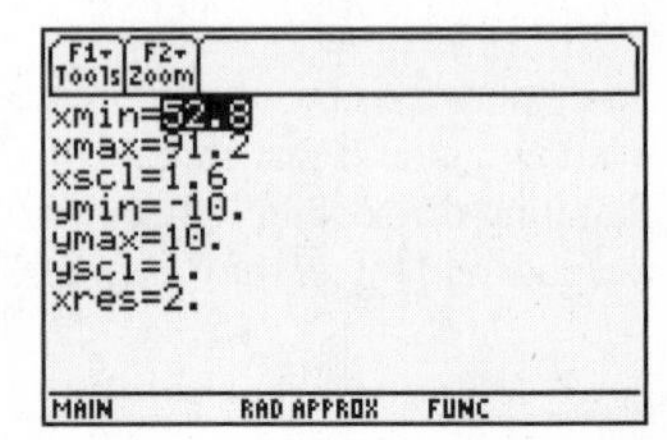

To force particular minimums, maximums and scaling we will press [◆][F2] (Window). This displays the screen at right. Notice the xmin was the smallest value shown on the plot; xmax is the largest. ymin and ymax are analogous. xscl and yscl are the distances between axis "tick marks." If your plot has lost the tops of the bars, you will need to increase ymax until they can be seen.

Change xmin to 55 (just slightly smaller than the smallest data value) ymin to -4 and ymax to 8. Ymin is set low so that the legends which appear after pressing Trace don't obscure portions of the graph, but don't set it so low that most of the window area is blank space.

To display the graph with the new window settings, press [◆][F3].

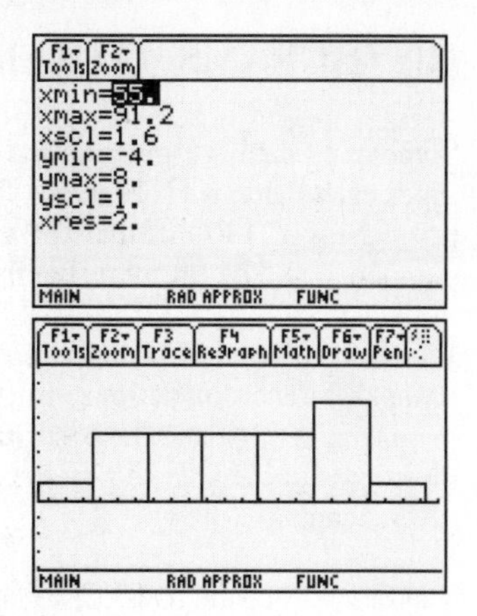

Manipulating the number of bars

The number of bars in a histogram is controlled by the value of Xscl on the 83/84 calculators and by the bucket width on the 89 series. In this graph, Xscl (the barwidth) was set to 3. The distribution looks unimodal again, but there is an interesting gap in the center. There are also gaps between the two smallest bars and the highest bar. It looks like we might have outliers.

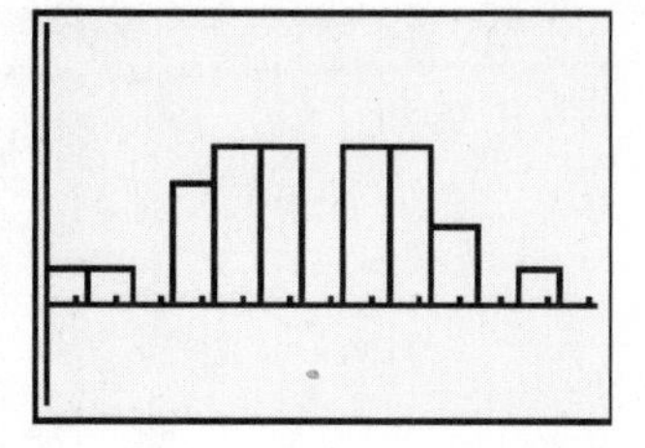

It's possible to have too few bars. Here, Xscl has been set to 10; Ymin is –3 and Ymax is 10. (These were changed for picture resolution.)

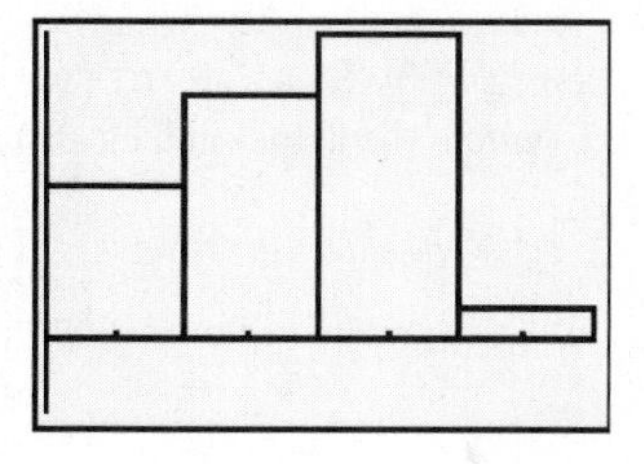

You will need to use your own judgment to decide how many bars to include and where. Your instructor may give you some guidelines. One rule of thumb for many years was to have somewhere between 5 and 20 bars; for most small data sets dividing the number of observations by 5 gives a good estimate of how many bars will give a decent picture.

"Printing" the Picture

Unfortunately, calculators do not have printers. To make a hard copy of the graph once you are satisfied, use the [TRACE] key to examine the entire graph. Make a picture of the histogram, clearly labeling each axis and giving the graph a title. Remember that the intervals given are the endpoints of the intervals. Label them as such. When you are finished, you should have a picture like the one at right. If you have TI-Connect software, you can use the screen capture application to save the picture on the computer for printing directly from the application, or use copy and paste to include the graph as a part of a word processing document. However, notice in the plots above that the screen capture does not include proper axis labels or titles!

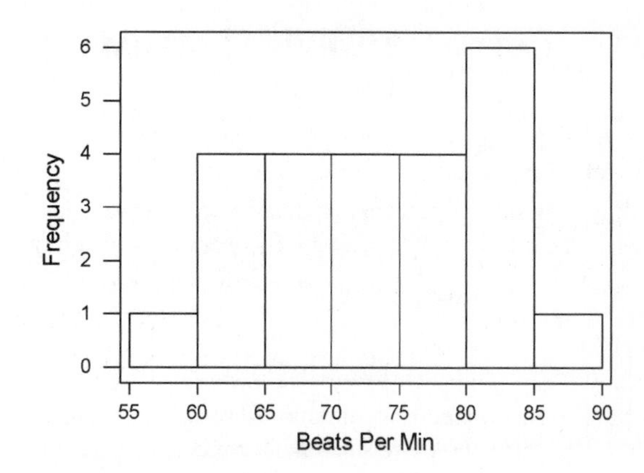

HISTOGRAMS WITH FREQUENCIES SPECIFIED

Height	Count
60	2
61	6
62	9
63	7
64	5
65	20
66	18
67	7
68	12
69	5
70	11
71	8
72	9
73	4
74	2
75	4
76	1

Sometimes data are given in the form of tables with both the data value and the number of times each value was observed. The frequency table on the side of the page shows the heights (in inches) of 130 members of a choir, as given in Problem 32 of Chapter 4. Entering 130 numbers could be tiresome, but there is a way to use the counts given.

We want to make a histogram to display this distribution. The procedure (allowing for the basic difference in defining the plot) is analogous on the 89 series calculators.) Enter the heights in one list and the observed counts in a second list. (We will put the heights in L1 and the counts in L2.) Make sure the lists are the same length, and that data values match with the given counts.

The first part of the lists looks like this.

L1	L2	L3
60	2	------
61	6	
62	9	
63	7	
64	5	
65	20	
66	18	

L1(1)=60

From [2nd][Y=] (Stat Plot) we will define Plot1 as at right. Notice Xlist is L1 (where the actual values are) and Freq is L2 (where the counts are.)

Pressing [ZOOM][9] gives the following graph. Notice that in this case the intervals and bar width (2) seem reasonable. This is again a right-skewed distribution (from the major peak, the right-hand side is longer than the left). Visually, the center is around 67 or 68 inches; from the data table the data range between 60 and76 inches.

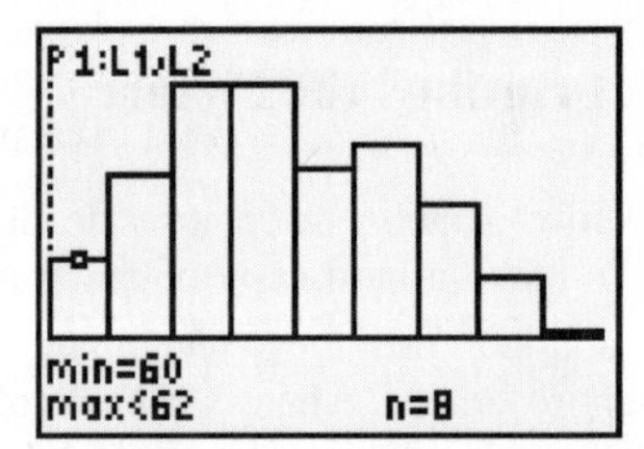

CALCULATING NUMERICAL SUMMARIES

Before computing any numeric summaries, we want to ensure that the calculator is using all possible decimal places in its intermediate calculations. It is best to use all digits possible, rounding any results only at the end. To do this, press [MODE]. The screen at right will be shown. The calculator should have Float highlighted. If not, move the cursor to highlight Float and press [ENTER] to move the highlight.

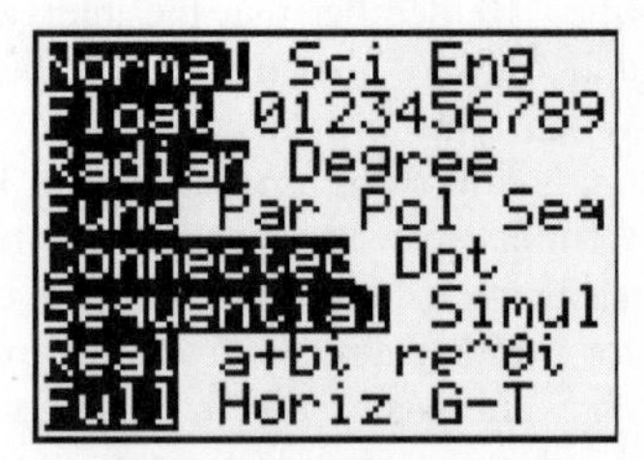

To calculate the numerical summary statistics for a single variable, first enter them into a list. Here the pulse rate data have been entered into list L1. The first few values are shown at right.

L1	L2	L3
88		------
56		
80		
60		
76		
72		
68		

L2(1)=

Press [STAT] then arrow to **CALC.** The menu at right will be displayed. The menu is organized so that the most frequently used options are at the top. Notice that **1:1-Var Stats** is highlighted. Press [ENTER] to select that option (or press [1]). The command will be transferred to the home screen.

Now you need to tell the calculator which list to use as input. Press [2nd][1] (**L1**). If no list name is given, the calculator will default to use **L1**, but it's good practice to specify the list name. Press [ENTER] to carry out the command.

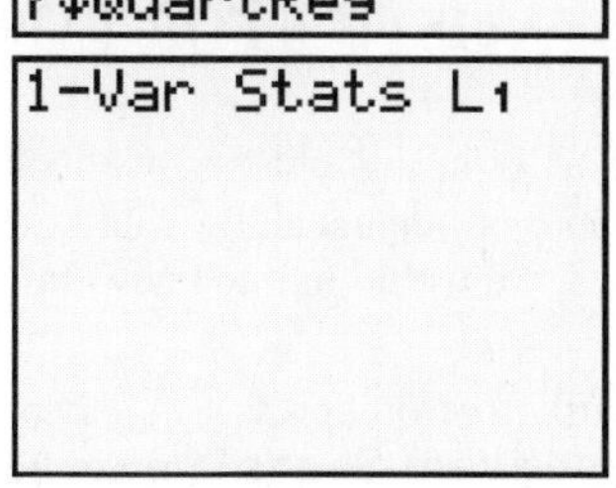

The first page of results is displayed at right. The arrow at the bottom left indicates more results are available and can be found by using [▼]. We first see the mean pulse rate is 72.833333. The calculator does not know if the data you are using represent a sample or a population. It has only one symbol for the mean ($\bar{x}$). If your data represents a population, you should report the mean using proper notation, and call it μ. The two values displayed next are the sum of the data values and the sum of squared data values. These are intermediate quantities used in computations, and are generally not of interest. Two different standard deviations are also reported as measures of spread.

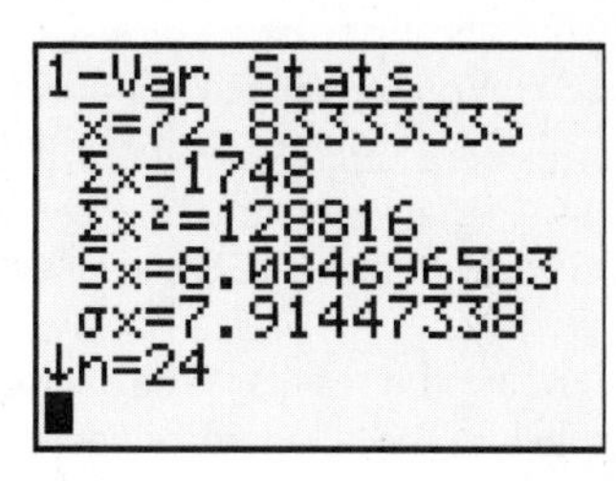

Sx $= \sqrt{\dfrac{\sum (x_i - \bar{x})^2}{(n-1)}}$ is the sample standard deviation and σx is the population standard deviation (the formula is almost the same; the divisor is n). This data does not represent all possible pulse rates even for the women studied, so we will use s_x (8.084696583) as the standard deviation. Which is the correct value to use depends on data you have – is it from a sample or is it for a population? The last value on the screen is the number of items in the data list; in this example, n is 24.

One thing to bear in mind is that calculators (and computers) will use (and report) many more digits than really make sense to use. It comes from division (in which, as we know, things don't always come out evenly) and taking square roots (which also aren't usually whole numbers). How many digits to report should be decided by your instructor, but a good rule of thumb is to report one more place than in the original data. Our data was in beats per minute, so we'll use one decimal place. Also, since we don't have all the possible pulse rates, we will report $\bar{x}$ = 72.8 and s = 8.1.

Using the down arrow, we find the five-number summary. The median (another measure of center) is 72, which is close to the mean in this data set (as it should be since the data were roughly symmetric – at least in our initial histogram). We can use the other values in this summary to compute two other measures of spread: the Interquartile Range (IQR) which is the spread of the middle half of the data, and the Range. The IQR is $Q_3 - Q_1$, or 80 - 68 which is 12. This means the central half of the data had a spread of 12 beats per minute. The Range is max – min, or 88 – 56 which is 32 beats per minute.

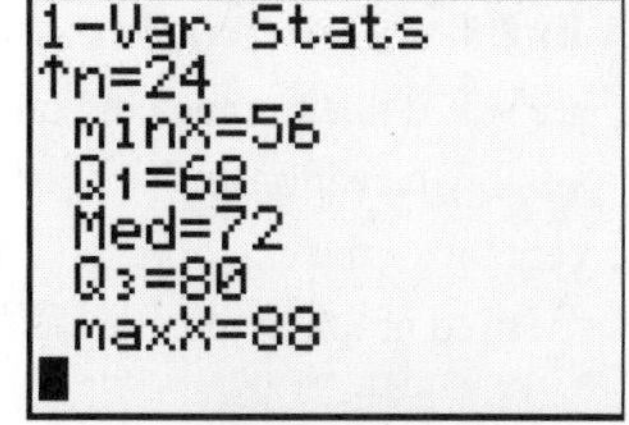

The procedure on a TI-89 is similar. Press [F4] (**Calc**). The menu at right will be displayed. The menu is organized so that the most often used options are at the top. Notice that **1:1-Var Stats** is highlighted. Press [ENTER] to select that option (or press [1]).

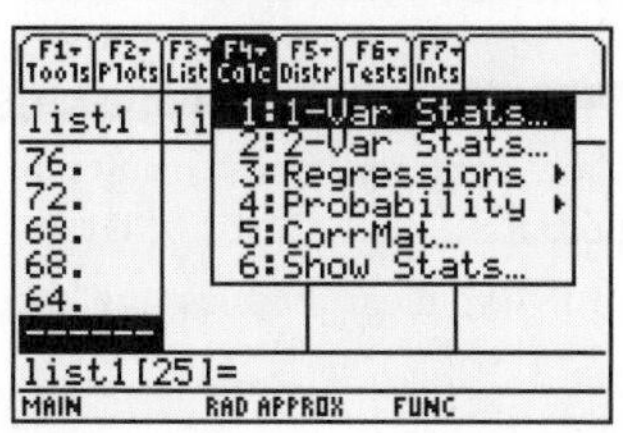

This is the input screen. You need to tell the calculator which list to use as input. Press [2nd][-] ([VAR-LINK]). Move the cursor to the correct list and press [ENTER] to select it. Since each value in our list occurred once, leave Freq at 1. Press [ENTER] to execute the command.

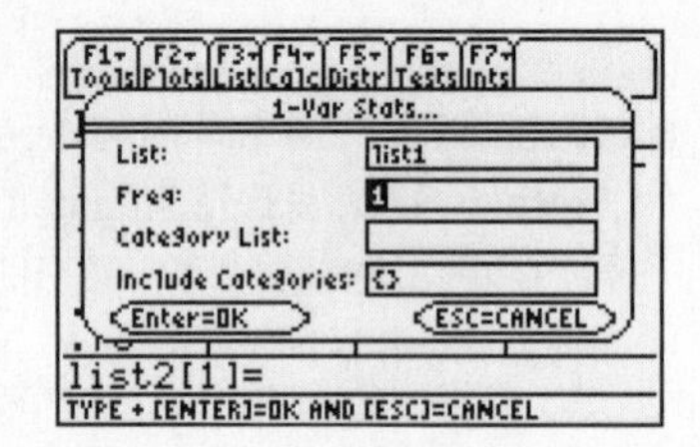

STATISTICS FOR TABULATED DATA

Earlier in this chapter, we looked at the distribution of heights of members in a choir. They were presented in a table of heights along with how many choir members there were of a given height. With these data in lists L2 and L3, we would like to know the average height for the choir.

Just as before, press [STAT] then arrow to CALC, press [ENTER] to select choice 1:1-Var Stats. On the home screen, you will specify not just one list, but two. The first list is the list of values (L2) and the second is the list of counts (L3). Your command will look like the screen at right. Don't worry that it didn't all fit on a single line. Press [ENTER] to execute the command. If you are using a TI-89 series calculator, proceed as in the example above, but set Use Freq and Categories to YES and specify the list of frequencies in the Freq box.

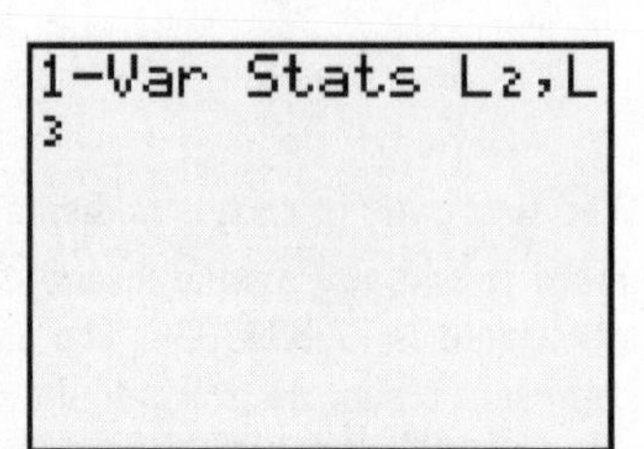

Here are the results. The average (mean) height for the choir members was 67.1 inches. The standard deviation (assuming these are not all the members possible for the choir) is 3.8 inches. If we were to consider this the entire membership of the choir, we would report $\sigma = 3.8$ (there is no practical difference between the two values here since n is large; the impact of subtracting 1 is not big.) Paging down, we find the median height was 66 inches. It is not surprising the median would be somewhat less than the mean for these data since the histogram indicated a right skewed distribution.

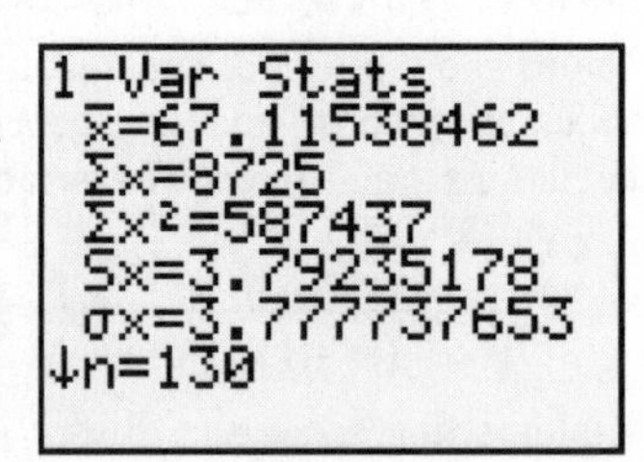

WHAT CAN GO WRONG?

Help! I can't see the picture!

Seeing something like this (or a blank screen) is an indication of a windowing problem. This is usually caused by pressing [GRAPH] using an old setting. Try pressing [ZOOM][9] to display the graph with the current data. This error can also be due to having failed to turn the plot "On."

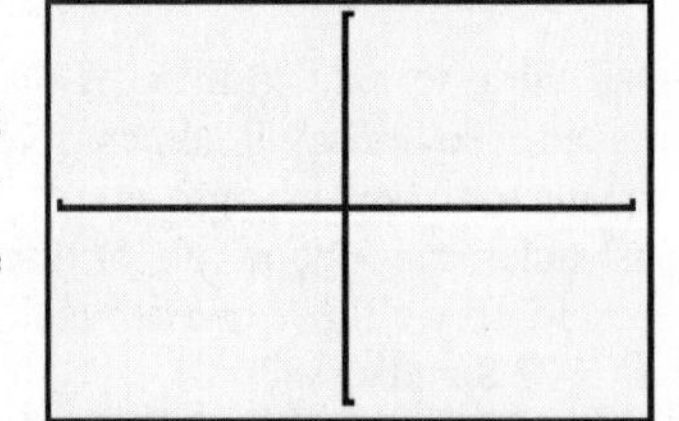

What's that weird line (or curve)?

There was a function entered on the [Y=] screen. The calculator graphs everything it possibly can at once. To eliminate the line, press [Y=]. For each function on the screen, move the cursor to the function and press [CLEAR] to erase it. Then redraw the desired graph by pressing [GRAPH].

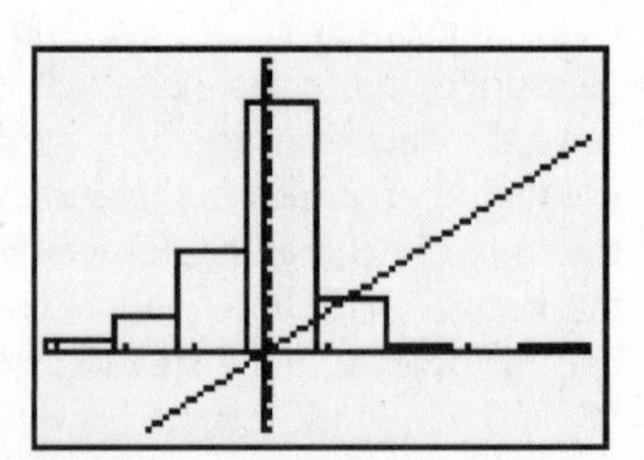

What's a Dim Mismatch?

This common error results from having two lists of unequal length. Here, it pertains either to a histogram with frequencies specified or a time plot. Press [ENTER] to clear the message, then return to the statistics editor and fix the problem.

What's an Invalid Dim?

ERR:INVALID DIM
1:Quit

This problem is generally caused by reference to an empty list. Check the statistics editor for the lists you intended to use, then go back to the plot definition screen and correct them.

What does Stat mean?

This error is caused by having two stat plots turned on at the same time. What happened is the calculator tried to graph both, but the scalings are incompatible. Go to the STAT PLOT menu and turn off any undesired plot.

Plot setup?

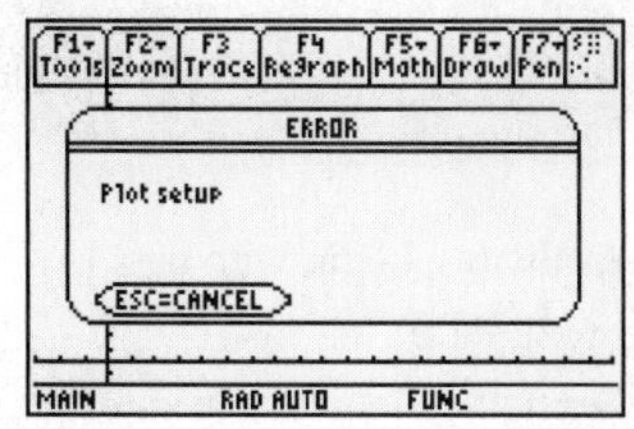

This is the TI-89 equivalent of the STAT error above. It is caused by having two stat plots turned on at the same time. The calculator tried to graph both plots, but the scalings are incompatible. Go to the Stat plots menu and turn off any undesired plot by moving the cursor to that plot, and pressing [F4].

Chapter 3 – Understanding and Comparing Distributions

In this chapter, we will meet a new statistics plot based on numerical summaries, a plot to track the changes in a data set through time, and ways to use plots to compare two or more distributions.

BOXPLOTS

Box plots (sometimes called box-and-whisker plots) are another way of picturing a distribution. Unlike histograms, they are based on definite values and are not subjective. However, as no plot is perfect, they can hide some potential features such as bimodality. A good practice, since it is generally so easy, is to look at several plots. They all can show different features of the distribution.

There are two types of boxplots – the original, which is based solely on the five-number summary (min, Q_1, median, Q_3, and max), and a "modified" boxplot, which has an objective criterion to identify outliers. Both types of plots divide the data into fourths – a "whisker" for the bottom and top quarters of the data, and a box for the middle half, with the median indicated inside the box. We always recommend using the modified boxplot, but your instructor may suggest otherwise.

As always, begin with data in a list. We will begin with the pulse rate data already examined in the last chapter. The data are in **L1**. Press [2nd][Y=] (**StatPlots**) then [ENTER] to get to the plot definitions screen for **Plot1**.

Notice there are two choices for boxplots - which is the boxplot where outliers are identified and which does not identify outliers. These are called **Box Plot** and **Mod Box Plot** on a TI-89. We will look at both to see the difference between the two.

Move the cursor to highlight . Make sure **Xlist** is changed to **L1** (press [2nd][1]); also make sure **Freq:** is set to 1 (press [ALPHA][1] if necessary). Your plot definition screen should look like this.

Press [ZOOM][9] ([F5] on an 89) to display the graph. The indications from the plot are that the distribution is (roughly) symmetric. Looking from the median line in the box to the two ends, the distances are relatively equal. However, since the median is somewhat to the left of the center of the box, one could call the distribution somewhat skewed. Pressing [TRACE] and using the right and left arrows will allow you to move around the graph, locating the median, quartiles, min, and max.

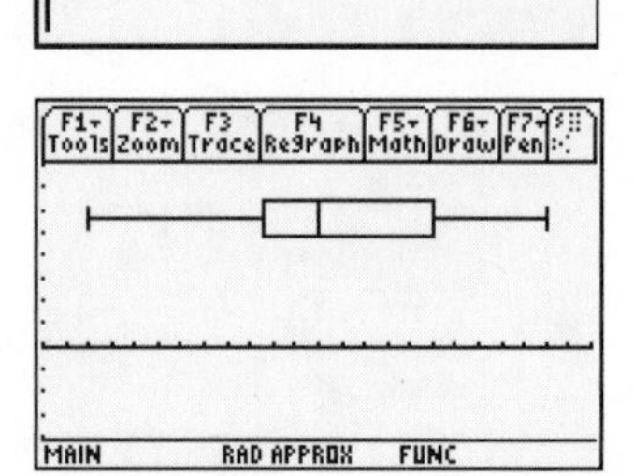

Return to the Plot definition screen and change the plot type to a boxplot identifying outliers ( ), or **Mod Box Plot** on an 89. Pressing [ZOOM][9] will display the plot at right. We see here that none of the values are outliers by the $Q_3 + 1.5*IQR$ and $Q_1 - 1.5*IQR$ criteria.

BOXPLOTS WITH TABULATED DATA

Reconsider the data on heights of members of a choir. According to the histogram, this was somewhat right-skewed. What will its boxplot look like? With heights in L2 and frequencies in L3, the plot definition screen looks like that at right.

After pressing [ZOOM][9] the graph at right will be displayed. Notice the median is not in the middle of the box; the right half of the data is longer than the left half (the data is right skewed) even though the two whiskers are relatively equal in length. Also, since we defined the plot to identify any possible outliers, none are flagged, so these data have no outliers.

HISTOGRAMS TO COMPARE DISTRIBUTIONS

We'd like to compare the distributions of two historically great baseball hitters: Babe Ruth and Mark McGwire. We have the following information on the numbers of home runs hit each year by Babe Ruth for 1920 through 1934 and for McGwire from 1986 to 2001.

Ruth:	52	59	35	41	46	25	47	60	54	46	49
46	41	34	22								
McGwire:	3	49	32	33	39	22	42	9	9	39	52
34	24	70	65	32	29						

[ZOOM][9] histograms as described in the previous chapter for both distributions are below.

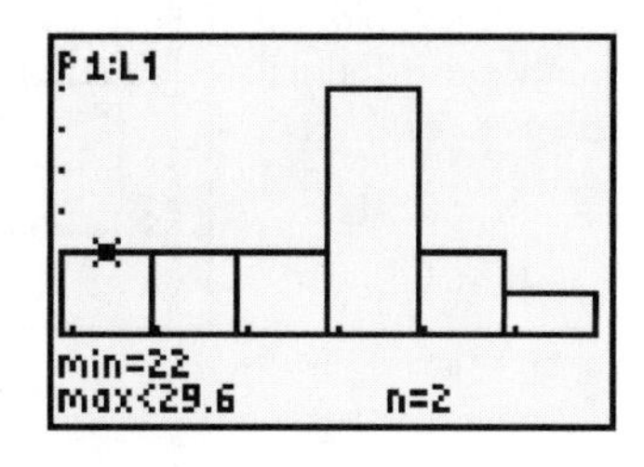

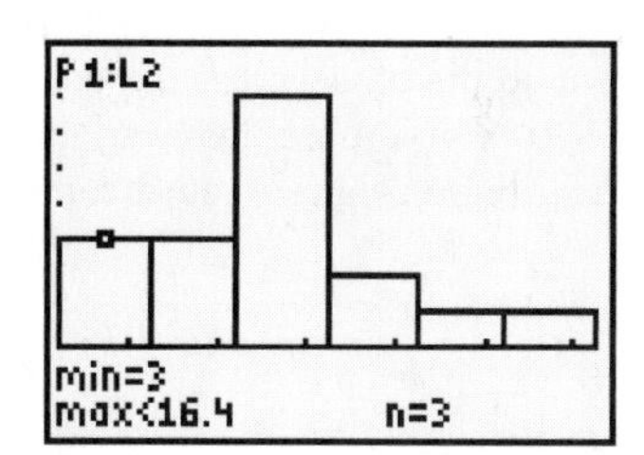

From the plots we can see both distributions are unimodal; Ruth's is skewed left and McGwire's is skewed right. But that's about all we can tell, because the graphs don't use the same scaling. People's eyes want to make a visual comparison, and since the graphs don't use the same values, this is impossible.

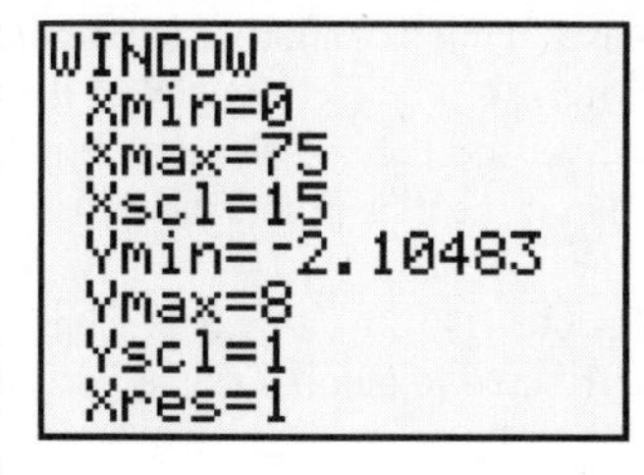

Let's change the scaling. Press [WINDOW]. Both the smallest and largest values occur in McGwire's distribution: 3 and 70. We also need a reasonable number of bars. It seems reasonable to set Xmin to 0, Xmax to 75 and use a bar width (Xscl) of 15. The settings used are at right, and the rescaled plots are below.

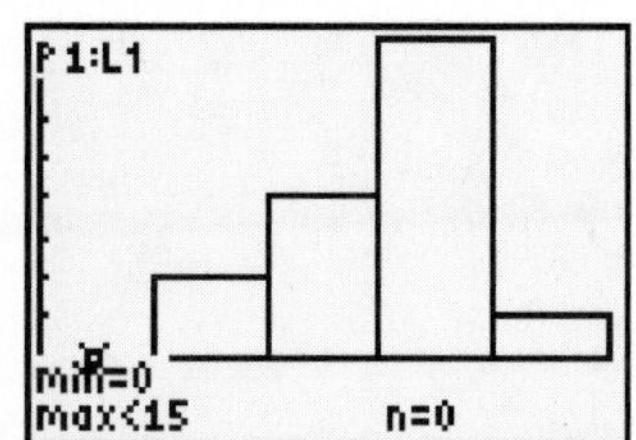

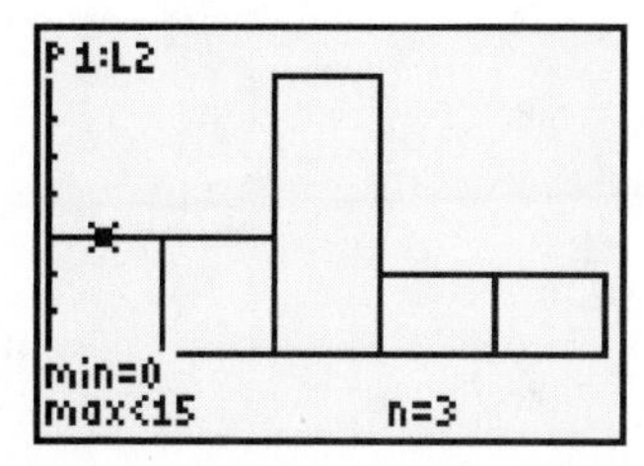

From these graphs it is easy to see that uth was the more prolific hitter. While McGwire had four years in which he hit more than 45 home runs, Ruth had 9. McGwire also had three seasons with fewer than 15 homers (due either to injury, strike, or his first, partial season). On this scaling, we still see Ruth's distribution as left-skewed, but McGwire's is fairly symmetric (even around the peak). Ruth's distribution has a smaller range (spread from low to high), and a higher center (visually about 60) than McGwire, whose center is between 45 and 60.

BOXPLOTS TO COMPARE DISTRIBUTIONS

Boxplots are very useful in comparing distributions. This is one of the few exceptions to the rule about only one plot being turned on at a time. Up to three boxplots can be displayed at once. Displaying boxplots side-by-side is a good way to make a visual comparison of distributions.

How would these distributions look displayed together as boxplots? The data have been entered into L1 (Ruth) and L2 (McGwire). We define Plot1 to use Ruth's home run values as at right.

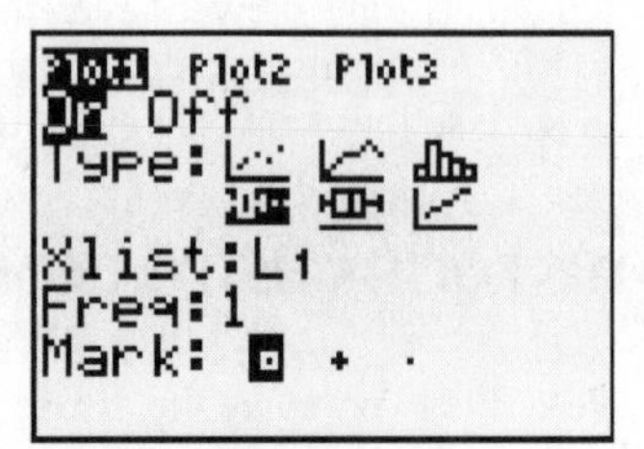

Returning to the STAT PLOT menu, we will arrow down to Plot2, and define it to use McGwire's numbers as at right.

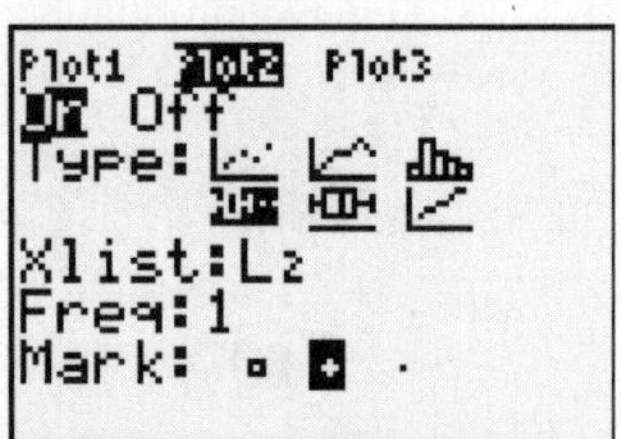

Pressing [ZOOM][9] gives the display at right. The top plot is Ruth's home run distribution; the bottom is McGwire's. (Pressing [TRACE] will identify each plot; to move from one to the other, press the down and up arrows.) Neither distribution has outliers. Ruth's is much less variable than McGwire's and is skewed left. McGwire's distribution appears rather symmetric.

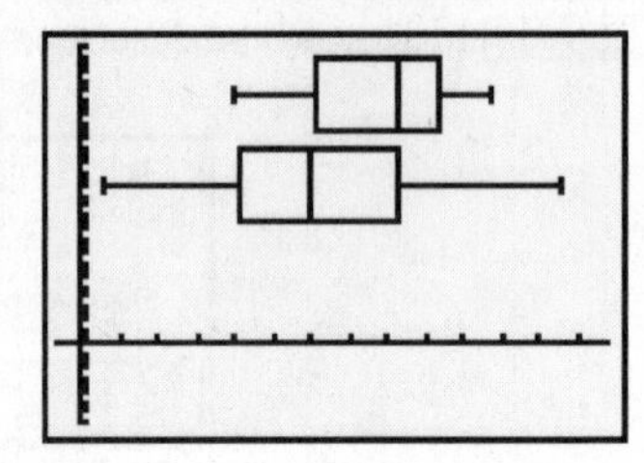

TIME PLOTS

Many variables are often measured at different points in time. It's not enough just to picture the distribution in this case. Time is an important factor, and we will want to know what (if any) part it plays. To answer this question, we will do a time (series) plot of the data. By convention, these are connected scatter plots with time represented on the *x*-axis and the actual variable values on the *y*-axis. They are connected because this makes any pattern easier to see than if the data points were just shown by themselves.

Problem 46 of the text in Chapter 5 is concerned with drunk driving. It lists the number of deaths (in thousands) attributed to alcohol-related driving form 1982 to 2005. The data are reproduced on the next page for convenience.

Year	Deaths	Year	Deaths
1982	26.2	1994	17.3
1982	24.6	1995	17.7
1984	24.8	1996	17.7
1985	23.2	1997	16.7
1986	25.0	1998	16.7
1987	24.1	1999	16.6
1988	23.8	2000	17.4
1989	22.4	2001	17.4
1990	22.6	2002	17.5
1991	20.2	2003	17.1
1992	18.3	2004	16.9
1993	17.9	2005	16.9

We could enter all the years manually, but it is easier to use the seq(command as described on page 10 of this manual. The data are in L2 and the years have been entered into L1.

Connected scatter plots are the second plot type on the plot definition screen. The Xlist is the time indices and Ylist is the number of drunk driving deaths. There is a choice of three options for marking the actual data points. You may pick whichever one you like; however, from past experience this author recommends the single pixel for this particular graph, as the others can make the plot look too cluttered. Once the plot has been defined, it can be displayed by pressing ZOOM 9.

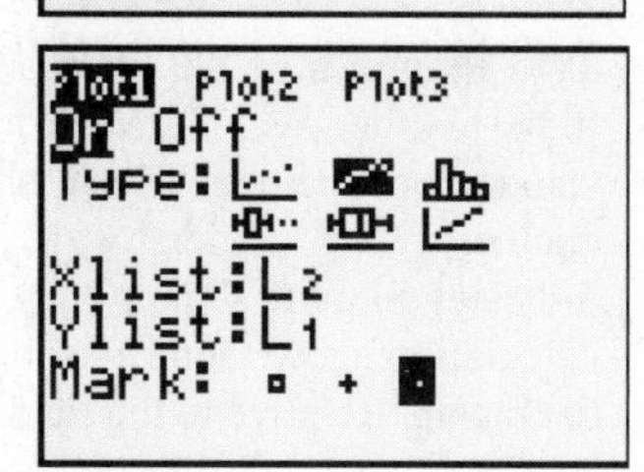

Here is the time plot. We can clearly a decline in alcohol-related fatalities in the early years of the data. However, the rate seems to have stabilized of late. What can be done to encourage less alcohol consumption combined with driving?

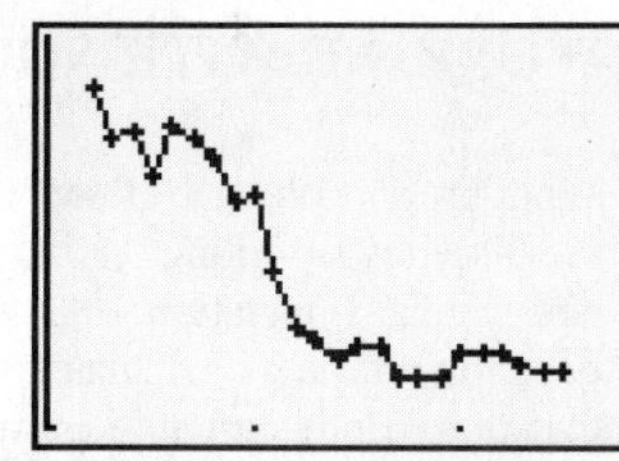

WHAT CAN GO WRONG?

Primarily, errors here are due to data entry mistakes. Always double check what you have entered. If a boxplot fails to display, the plot may not have been turned on. Always be sure to turn off any extra plots after copying them to paper. If not, you probably will receive either the Invalid Dim (referring an empty list) or Stat(Plot Setup on a TI-89) which is the incompatible window ranges for two plots error message. These were discussed in the previous chapter.

Chapter 4 – The Standard Deviation as a Ruler and the Normal Model

The standard deviation is the most common measure of variation; it plays a crucial role in how we look at data. Z-scores measure standard deviations above or below the mean as a pure number (no units) and are useful as measures of relative standing. Normal models are very useful as many random variables (at least approximately) follow their unimodal, symmetric shape.

Z-SCORES

The *z*-score for an observation is $z = \frac{(obs - \overline{y})}{s}$ or $\frac{obs - \mu}{\sigma}$ for a population, where *obs* is the value of interest. Positive values indicate the observation is above the mean; negatives mean the value is below the mean. Calculating them is easy as long as one keeps in mind that the subtraction in the numerator must be done before the division. Calculators follow the arithmetic hierarchy of operations.

For example, in the 2004 Olympic women's heptathlon, Austra Skujyte of Lithuania put the shot 16.4 meters; the mean distance for all shotputters in the contest was 13.29 m with standard deviation 1.24 m. Carolina Kluft won the long jump with a 6.78 m jump; the average for all contestants was 6.16 m with standard deviation 0.23 m. Who actually did "better" relative to the other contestants? Scaling makes a direct comparison of the performances difficult, if not impossible. Z-scores can answer the question.

Two examples of the calculation for Skujyte are at right. The first (incorrect!) indicates she was 5.7 standard deviations *above* the mean – unreasonable given the standard deviation. The problem is failing to perform the subtraction first or enclosing the numerator in parentheses. The second (correct!) calculation indicates Skujyte's jump was 2.51 standard deviations above the mean. The last calculation shown indicates that Kluft's jump was 2.70 standard deviations above the mean. Relative to the field, her jump was the better performance.

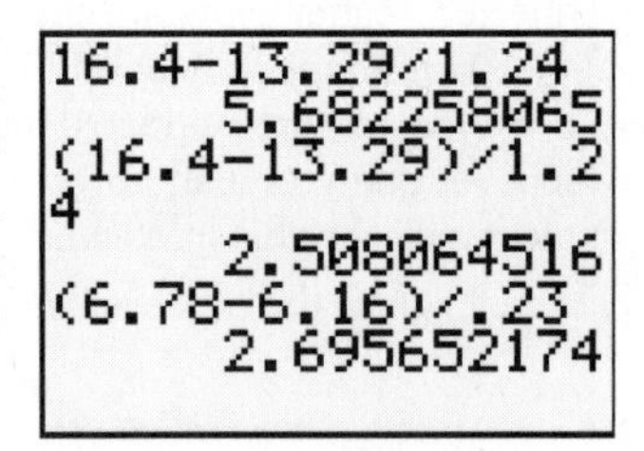

WORKING WITH NORMAL CURVES

What proportion of SAT scores are between 450 and 600? SAT scores for each of the three tests (writing, verbal and math) are approximately normal with mean 500 and standard deviation 100, or N(500, 100). There are two ways to answer this question with a TI calculator. One will draw the curve; the other just answers the question. Both start at the same place: the Distributions menu. On an 83/84, press [2nd][VARS] (`DISTR`). If you are using an 89, the `Distr` Menu is [F5] in the `Statistics/List Editor` application. On a TI-83/84, the screen at right will appear. Notice the arrow pointing down at the bottom left. There are more distributions which can be used; more will be said about some of them later. At the top of the screen, there are two choices `DISTR` (the default) which merely gives distribution values, and `DRAW` which will draw the curves and shade the appropriate areas.

First, let's answer the question. On an 83/84 the menu option to select is `2:normalcdf(`; `Normal Cdf` is menu option 4 on 89s. Either press the down arrow and then [ENTER] or just press [2]. The command will be transferred to the home screen. It normally requires two parameters to be entered: the *z*-score for the low end of the area of interest and the *z*-score for high end. Separate the entries by commas, and finish by closing the parentheses.

A score of 450 is 0.5 standard deviations below the mean, so its *z*-score is –0.5. A score of 600 is 1 standard deviation above the mean; its *z*-score is 1.

The command has been entered in the screen at right, and ENTER was pressed to execute the command. We see that about 53.3% of all scores on the SAT will be between 450 and 600. TI-83/84 calculators default to a mean of 0 and standard deviation 1. Since we are working with standardized values (*z*-scores), there is no need to input these values.

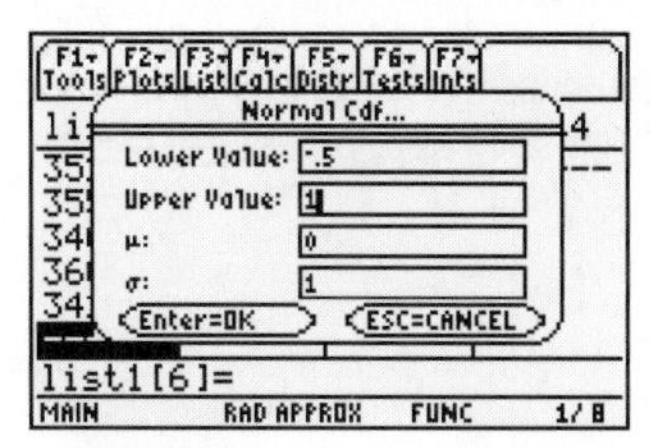

At right is the TI-89 series input screen. With these calculators, you need to explicitly specify the mean and standard deviation for the distribution of interest.

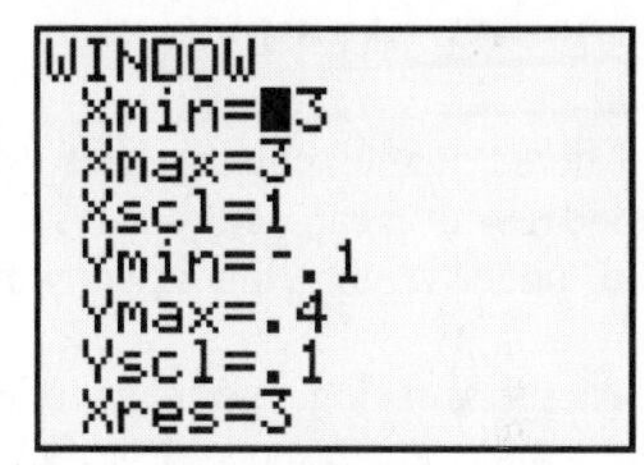

With TI-83/84 calculators, to find the area and have it shaded, one needs to first set the window (ZOOM 9 does not work here). For any normal model, most values will range from about –3 to 3 (three standard deviations either side of the mean) by the 68-95-99.7 Rule. Since the whole area under the curve is 1, the height of the curve will be a small number; we have set the `Ymin` to –0.1 and `Ymax` to 0.4 as in the screen at right.

Now press 2nd VARS (**DISTR**) and arrow to **DRAW**. We want choice 1, so press ENTER.

The command has been transferred to the home screen. Enter the *z*-scores for the low end of interest and the high end as at right, then press ENTER.

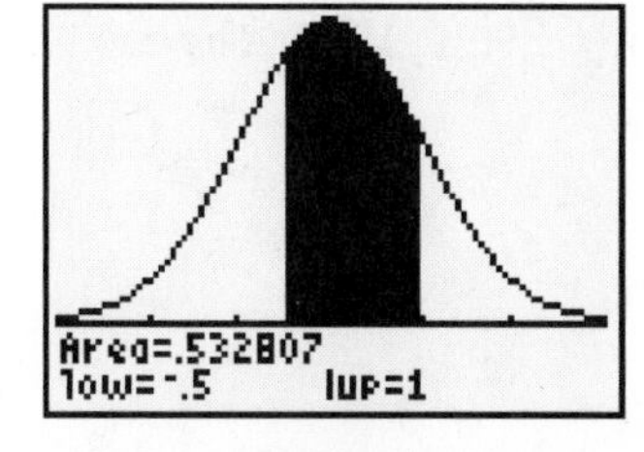

The graph should look like the one at right. Again, we see the area is about 53.3%; we also see what portion of the normal curve it represents.

When working with the **DRAW** option, the graph must be cleared between successive commands or the shaded area will accumulate until the whole curve is shaded. To clear the drawing, press 2nd PRGM (**DRAW**) then press ENTER to select option `1:ClrDraw`.

If you are using a TI-89 series calculator, this is a little more straightforward. From the F5 `Distr` menu, press the right arrow key to expand the **Shade** menu. `Shade Normal` will be highlighted. Press ENTER to select that option.

The input screen is similar to the one seen above that simply computes the area. There is one major difference at the bottom. You have the option to have the calculator `Auto-scale` the graph. Use the right arrow key to expand the option box, and the down arrow key to move the selection to YES, if needed.

Displaying the graph puts you into the `Graph` application. To return to the Statistics app, press [2nd][APPS].

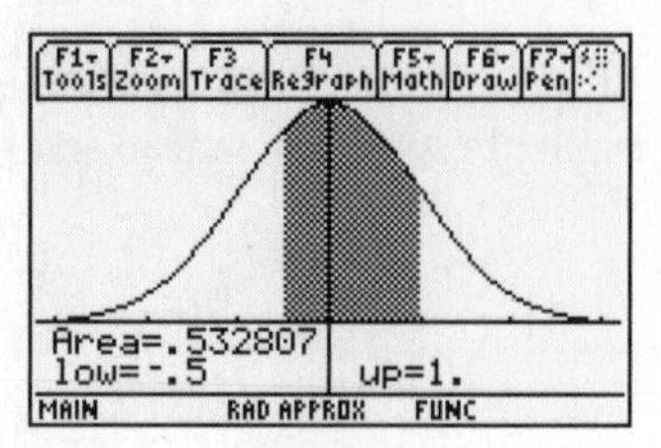

Another Example

A cereal manufacturer makes boxes labeled as 16 ounces; but the boxes are actually filled according to a normal model with mean 16.3 ounces and standard deviation 0.2 ounces. We want to know what fraction of all boxes will be "underweight," that is, contain less than the advertised 16 ounces.

Strictly speaking, Normal models extend from -∞ to ∞ (negative infinity to infinity). On the calculator, ∞ is represented as 1e99 (10^{99}). To enter this, one presses [1][2nd][,][9][9], but practically, any "very large" negative number (say, -99) will work for -∞ and any large positive number (say 99) for ∞ since we know almost all of the area is between –3 and 3 standard deviations away from the mean.

We want to know what fraction of all boxes are less than 16 ounces, so the low end of interest is -∞ (we entered –99 as the stand-in); the upper end of interest is 16 which corresponds to a *z*-score of –1.5. The command and the result are at right. We see that about 6.7% of all boxes of this cereal should be underweight.

WORKING WITH NORMAL PERCENTILES

Sometimes the area under the curve is given and the corresponding value of the variable is of interest. For example, in the SAT model used before, how high must a student score to place in the top 10%? In a sketch of the normal curve, the unknown value, we'll call it *X*, separates the top 10% from the lower 90%. We first have to find a corresponding *z*-score.

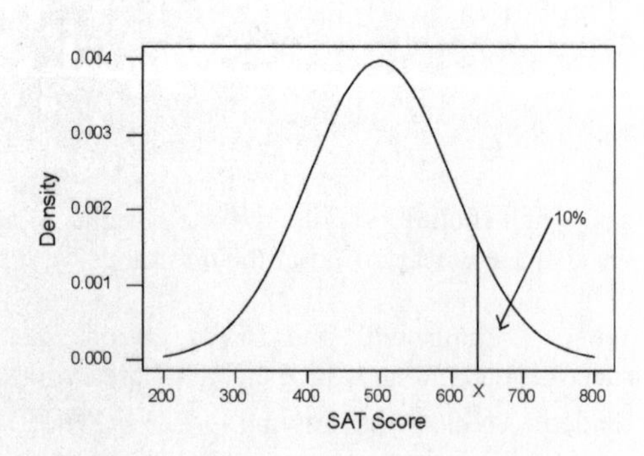

This is the opposite, or inverse, situation of that we've just explored. On the TI-83/84 `DISTR` menu, the command is `3:invNorm(`. Press [2nd][VARS][3] to transfer the command to the home screen. The parameter for this command is area to the *left* of the point of interest (.90 or 90%). Press [ENTER] to execute the command. The *z*-score of interest is 1.28. To be in the top 10%, your score must be 1.28 standard deviations above the mean. We have to solve the equation

$z = (x - \mu)/\sigma = 1.28$ or $(x - 500)/100 = 1.28$ After doing the algebra, we see that a score of 628 will put a person in the top 10% of all SAT scores; practically since scores are reported rounded to multiples of 10, a score of 630 is needed.

If you are using a TI-89, use `Inverse Normal` from the [F5] Distr menu, found by expanding option `2:Inverse`.

To do this computation on TI-89 calculators, you must again specify the mean and standard deviation (0 and 1) as I have done here. You will find the same *z*-score as above. Complete the algebra to find the SAT score of interest.

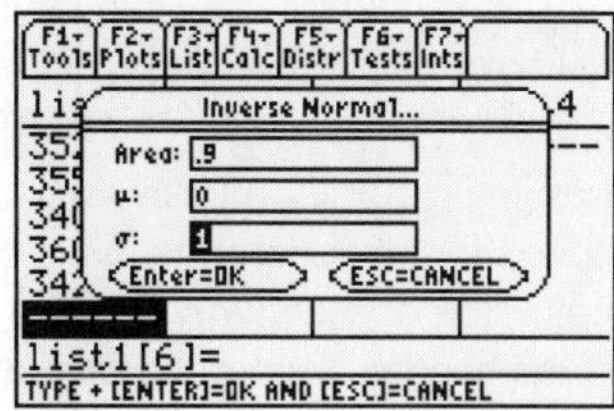

Another Example

The cereal company's lawyers are not happy with 6.7% of boxes being underweight. They want at most 4% to be underweight. What mean must the company reset its machines to in order to achieve this target? We'll use `invNorm` for a standard normal model to find the *z*-score corresponding to 4% of the area below this value, then use some algebra to solve for the unknown mean.

The *z*-score of interest is –1.75. We need to solve the equations

$z = (x - \mu)/\sigma = -1.75$

$(16 - \mu)/0.2 = -1.75$

Now multiplying both sides by 0.2, and subtracting 16 from both sides gives $-\mu = -16.35\sigma_X^2$ or μ = 16.35. In order to achieve the target of no more than 4% of boxes being underweight, the machine will have to be set for an average of 16.35 ounces per box.

```
invNorm(.04)
        -1.750686071
Ans*.2-16
        -16.35013721
```

Now, suppose the president of the cereal company wants the mean to be no more than 16.2 oz (she doesn't want to give away too much cereal.) To meet that target along with the 4% target for the proportion of boxes that are underweight, the company must change the standard deviation. We already know that a z-score of -1.75 corresponds to the 4th percentile. What changes here is the algebra: we know the desired μ and want to find σ. $z = (x - \mu)/\sigma = -1.75$ becomes $(16 - 16.2)/\sigma = -.2/\sigma = -1.75$, and finally we find $\sigma = 0.114$ oz.

IS MY DATA NORMAL?

It is one thing to assume data follows a normal model. When one actually has data the model should be verified. One method is to look at a histogram: is it unimodal, symmetric and bell-shaped? Another method is to ask whether the data (roughly) follow the 68-95-99.7 rule. Both of these methods might work well with a fairly large data set; however, there is a specialized tool called a normal probability (or quantile) plot that will work with any size data set. This plots the data on one axis against the *z*-score one would expect if the data were exactly normal on the other. If the data are from a normal distribution this plot will look like a diagonal straight line.

TI-83/84 Procedure

Recall the data on pulse rates. In one histogram, they appeared symmetric and unimodal. In another, they appeared somewhat uniform. Might we consider these data as having come from a normal model? The data are in list L1. Press [2nd][Y=] to get to the first Stat Plot screen. Once here, you should always check that all plots are off except the one you will use. Select the plot to use and press [ENTER]. The normal probability plot is the last plot type. Use the right arrow to move there and press [ENTER] to move the highlight. Notice you have a choice of having the data on either the x or y axis. It doesn't really matter which you choose; many statistical packages put the data on the x-axis; many texts (including DeVeaux, Velleman, and Bock) put the data on the y-axis. As we have seen before, you have a choice of three marks for each data point. Select the one you prefer.

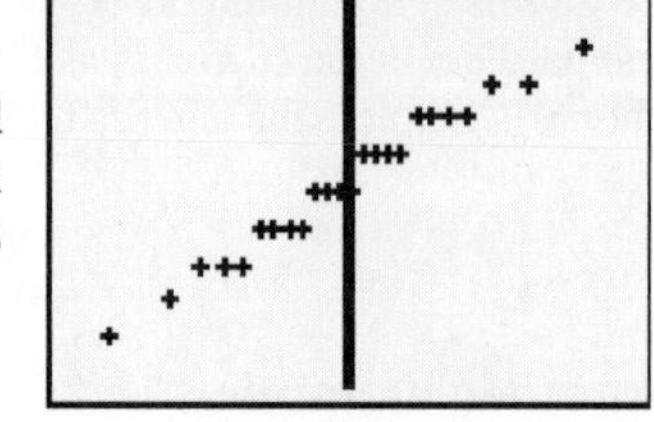

Pressing [ZOOM][9] displays the graph. This graph is very straight. These data could indeed be considered to have come from a normal distribution. We also see an indication of *granularity* in the plot. That is because we have several data values which occurred several times.

TI-89 Procedure

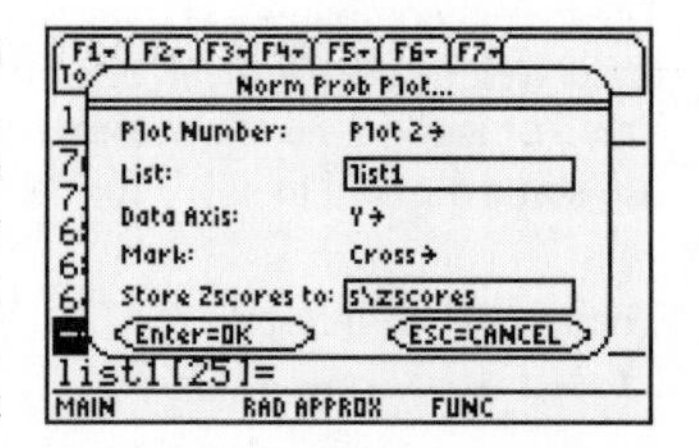

With the pulse rate data in `list1`, press [F2] (`Plots`) and select choice `2:Norm Prob Plot`. The plot number defaults to one more than the last plot defined. Enter the list name to use (press [2nd][-] to get to the `VAR-LINK` screen). Notice you have a choice of having the data on either the x- or y-axis. It doesn't really matter which you choose; many statistical packages put the data on the x-axis; many texts (including DeVeaux, Velleman and Bock) put the data on the y-axis. As we have seen before, you have a choice of marks for each data point. Select the one you prefer. The calculator will store the z-scores in a list. Just take the default here. Pressing [ENTER] calculates the z-score for each data value. To display the plot, press [F2] again, check that all other plots are "turned off" (uncheck them by moving the cursor and pressing [F4]) then press [F5] to display the plot.

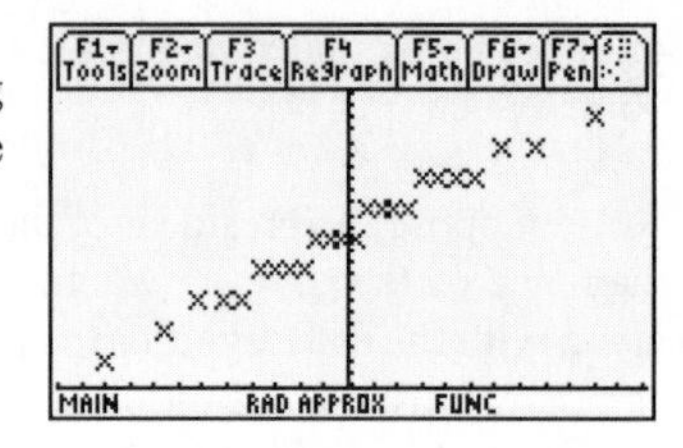

The graph is very close to a straight line. These data may be considered as having come from a normal distribution. We also see an indication of *granularity* in the plot. That is because we have several data values which occurred several times.

Skewed distributions often show a curved shape. Data on the cost per minute of phone calls as advertised by Net2Phone in USA Today (July 9 2001) to 22 countries were as follows:

7.9	17	3.9	9.9	15	9.9	7.9	7.9	7.9	7.9	8.9
21	6.9	11	9.9	9.9	7.9	3.9	22	9.9	7.9	16

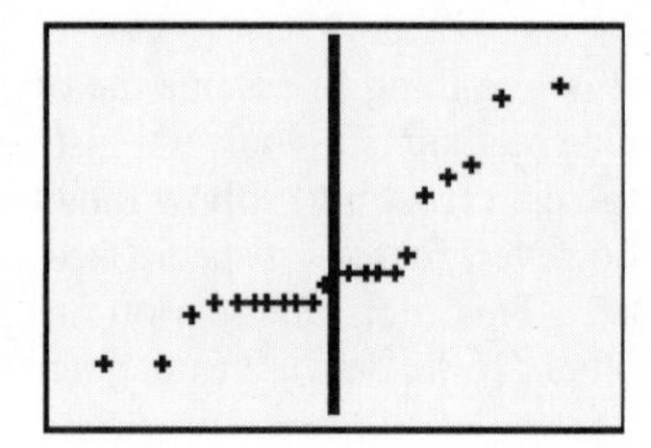

We have entered in a list and have defined the normal probability plot as above.

The plot obtained is at right. Not only does it show a general upward curve, it also displays granularity again. This occurs when a particular data value occurs several times (as with 7.9 cents per minute which was in the list seven times.)

WHAT CAN GO WRONG?

Why is my curve all black?

For the SAT scores curve, the graph indicates more than half of the area is of interest between 200 and 475 (z-scores of –3 and -.25); the message at the bottom says the area is 40%. This is a result of having failed to clear the drawing between commands. Press [2nd][PRGM] then [ENTER] to clear the drawing, then reexecute the command.

```
Area=.399944
low=-3        up=-.25
```

How can the probability be more than 1?

It can't. If the results look like the probability is more than one, check the right side of the result for an exponent. Here it is –4. That means the leading 2 is really in the fourth decimal place, so the probability is 0.0002. The chance a variable is more than 3.5 standard deviations above the mean (this would be a box of the cereal more than 17 ounces) is about 0.02%.

```
normalcdf(3.5,99
)
   2.326733735E-4
```

How can the probability be negative?

It can't. The low and high ends of the area of interest have been entered in the wrong order. As the calculator does a numerical integration to find the answer, it doesn't care. You should.

```
normalcdf(1,-99)
      -.8413447404
```

What's Err: Domain?

This message comes as a result of having entered the `invNorm` command with parameter 90. (You wanted to find the value that puts you into the top 10% of SAT scores, so 90% of the area is to the left of the desired value.) The percentage must be entered as a decimal number. Re-enter the command with parameter .90.

Chapter 5 – Scatterplots, Correlation, and Regression

Are two numeric variables related? If so, how? Scatterplots and regression will answer these questions. Correlation describes the direction and strength of linear relationships. Linear regression further describes these relationships.

Here are advertised horsepower ratings and expected highway gas mileage for several model year 2007 vehicles. These are the data for problem 35 in Chapter 7 of the text.

Audi A4	200 hp	32 mpg	Honda Accord	166	34
BMW 328	230	30	Hyundai Elantra	138	36
Buick LaCrosse	200	30	Lexus IS 350	306	28
Chevy Cobalt	148	32	Lincoln Navigator	300	18
Chevy Trailblazer	291	22	Mazda Tribute	212	25
Ford Expedition	300	20	Toyota Camry	158	34
GMC Yukon	295	21	VW Beetle	150	30
Honda Civic	140	40			

How is horsepower related to gas mileage? The first step in examining relationships is through a scatterplot.

SCATTERPLOTS

Here are the first few values for the horsepower ratings (in `L1`) and the gas mileages (in `L2`). It is important to enter these very carefully as they have been entered in the table, since the values represent a data pair for each vehicle type.

Our supposition is that larger engines will get less gas mileage, so we will use the horsepower ratings as the predictor (*X*) variable and the gas mileage as the response (*Y*) variable.

To define the scatterplot, press [2nd][Y=] (`STAT PLOT`). Select `Plot1` by pressing [ENTER]. Scatterplots are the first plot type. Move the cursor to highlight that plot, and press [ENTER] to move the highlight. Press the down arrow ([▼]) and enter the list where the predictor (*X*) variable is (here, `L1`, so [2nd][1]). Press the down arrow and enter the list containing the response (*Y*) variable (here, `L2`, so [2nd][2]). Press [▼] to select the type of mark for each data point (*X*, *Y*) pair. The author of this manual recommends either the square or cross; the single pixel tends to be too hard to see. When finished, the plot definition screen should look like the one at right.

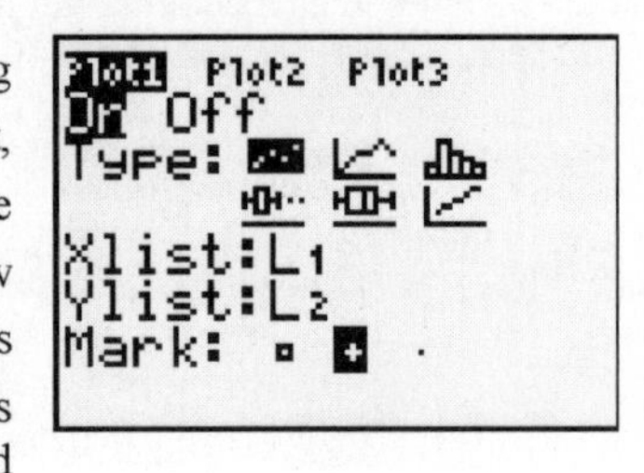

To define the scatterplot on a TI-89, press [F2] (`Plots`) then select choice `1:Plot Setup`. Make sure unnecessary plots are cleared out by moving the cursor to highlight the plot and pressing [F3] to clear the plot or [F4] to uncheck it. Press [F1] to start defining `Plot 1`. Press the right arrow to select the plot type. Press [ENTER] to select `1:Scatter`. Press the down arrow ([▼]) and select your choice of `Mark` for the data points by using the right arrow to expand the menu, then the down arrow to make your selection. The author recommends against the last choice `5:Dot` as the single pixel tends to be too hard to see. Press the down arrow and select the list with the predictor (*X*) values ([2nd][-] gets the [VAR-LINK]screen; move the cursor to the desired list and press [ENTER]). Press the down arrow and enter the list containing the response (*Y*). When finished, the plot definition screen should look like the one at right.

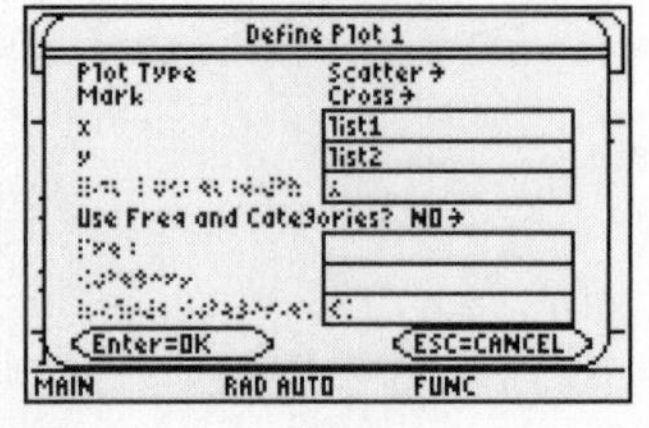

Press [ZOOM][9] ([F5] on an 89) to view the plot. Here we see a generally decreasing pattern from left to right, supporting our initial idea. The pattern is generally linear; however, some points at the bottom right may be unusual; we'll examine those later.

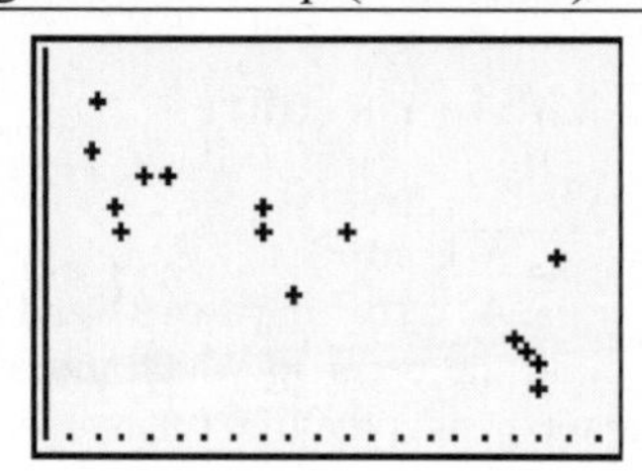

REGRESSION SETUP (TI-83/84 ONLY)

We want to examine the nature of this relationship further; before we do, we need to set up the calculator to display the values of the correlation coefficient (r) and the coefficient of determination (r^2). This only needs to be done if you are using a TI-83/84. The TI-89 series will always display these quantities.

This procedure *normally* needs to be done only *once*; however, changing batteries slowly will reset memory and it may have to be done again.

Press [2nd][0] (`Catalog`). This accesses the list of all the functions the calculator knows about. Notice that the cursor is at the beginning of the catalog. We need to get down to a command that begins with a D, so press [x^{-1}] (The equivalent of alpha D. The cursor is already in Alphabetic mode, indicated by the A at the upper right on the screen).

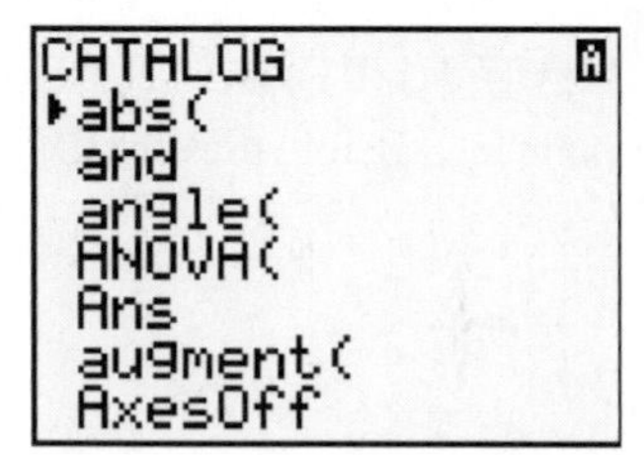

We're now here. We haven't gotten to the command yet, but we're close.

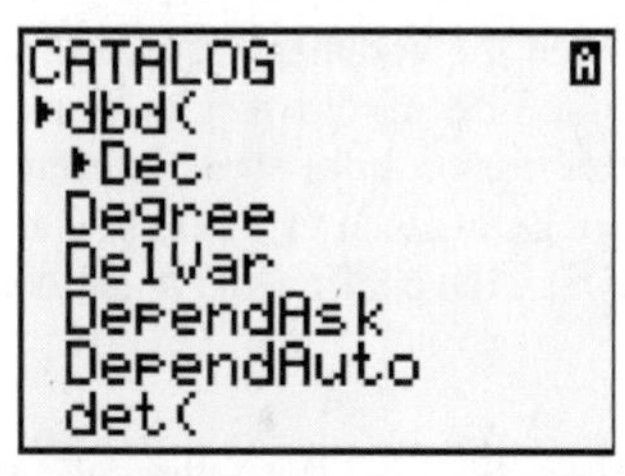

Press the down arrow [▾] until the command `DiagnosticOn` is highlighted. Press [ENTER] to select the command and transfer it to the home screen, then [ENTER] again to execute it.

Your screen should look like the one at right.

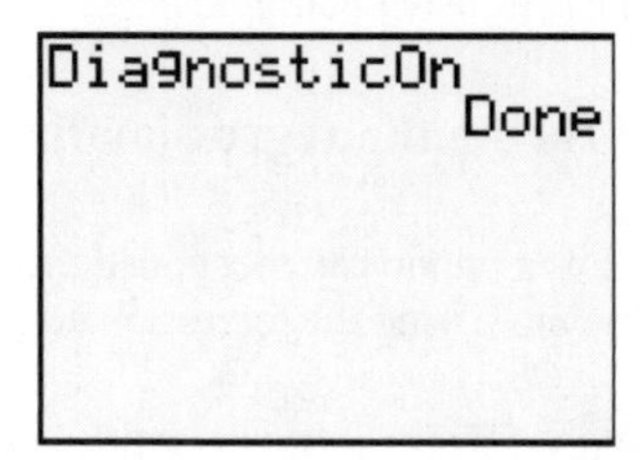

REGRESSION AND CORRELATION

We're now ready to examine the correlation between these two variables. However, the calculator will not give just the value of r; it's much easier computationally for it to do the whole thing at once and report all the statistics of interest.

TI-83/84 Procedure

Press [STAT], arrow to CALC. (We've been here before for 1-var Stats). There are two linear regression choices: 4:LinReg(ax+b) and 8:LinReg(a+bx). The answers you get will be the same, but one must keep in mind the order in which the coefficients are used. Since statisticians usually prefer the constant term of the regression to come first (in case there are several predictor variables – multiple regression) we'll use choice 8.

Either press the down arrow until the selection is highlighted then [ENTER] or simply press [8]. The command will be transferred to the home screen. In doing a simple regression with predictor variable in L1 and response in L2, simply pressing [ENTER] at this point is enough. However, if you want to store the equation (to see it on your graph for one reason), you need to specify the lists in which the data is stored; it's also just good practice to get into the habit, since you may want to use lists other than L1 and L2.

Press [2nd][1] (L1), then [,], then [2nd][2] (L2) followed by [ENTER] to execute the command. Before pressing [ENTER] the screen should look like the one at right.

Once the command is executed, you should see the display at right. Notice the first line of the results displays the type of regression in terms of y and x. Regression lines should never be reported in these terms, but the calculator does not know what variables you are working with. This is really an aid to remind you where the coefficients a and b go in the equation.

Here, we have the following regression equation: *EstimatedMileage* = 46.87 − 0.08**horsepower*. As always, how many decimal places to report can be subjective (and depend on the type of the data). For data like these, two decimal places should be sufficient. Ask your instructor for guidance. Remember, this line represents the average value of gas mileage for a given horsepower rating, based on the model from our data. The slope of -0.08 indicates that gas mileage decreases about 0.08 miles per gallon for each additional horsepower in the engine, on the average. In addition, we see the correlation coefficient, $r = -0.869$ which indicates a strong negative relationship. The coefficient of determination, $r^2 = 0.755$ (normally expressed as $r^2 = 75.5\%$) tells us that 75.5% of the observed variation in gas mileage (remember these values ranged from 10 to 31 mpg) is explained by the horsepower of the engine, using our model.

Storing the regression line

It was previously mentioned that the equation of the regression line can be stored for future reference. This is done by modifying the regression command as follows:

Press [STAT], arrow to CALC, then [8][2nd][1][,][2nd][2] (so far, this is what we did before). Now press [,][VARS] arrow to Y-Vars, press [ENTER] to select 1:Function, then [ENTER] to select Y_1. The regression command should look like that at right. Press [ENTER] to execute the command.

The regression output will look the same as before. The difference between the two commands can be seen by pressing [Y=]. The equation of the line has been stored for further use.

It would be nice to see how the line passes through the data; it should be roughly in the center of the data points. Press [GRAPH], since there is no need to resize the window. Sure enough, there's the line just as we expected. Notice that since no line will be perfect (unless $r = \pm 1$), some of the points are above the line, and some below. The vertical distances between the points and the regression line are called *residuals* and their plots are used to examine the line for adequacy of the model.

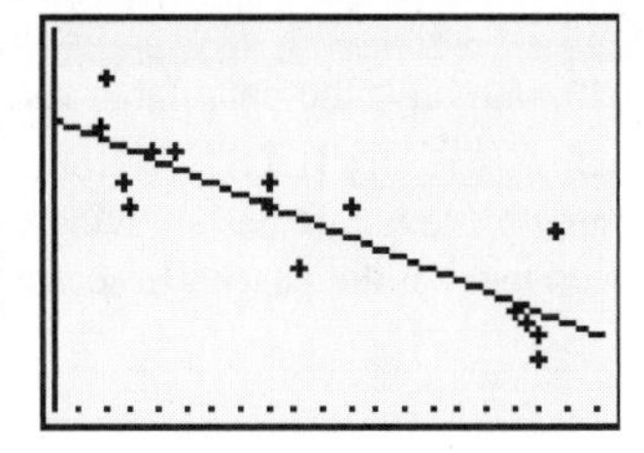

TI-89 Procedure

From the Statistics Editor, press [F4] (`Calc`). Arrow down to choice `3:Regressions` and press the right arrow. Both choices 1 and 2 on this submenu are linear regressions. The answers you get will be the same, but one must keep in mind the order in which the coefficients are used. Since statisticians usually prefer the constant term of the regression to come first (in case there are several predictor variables – multiple regression) we'll use choice 1. Press [ENTER] to select it.

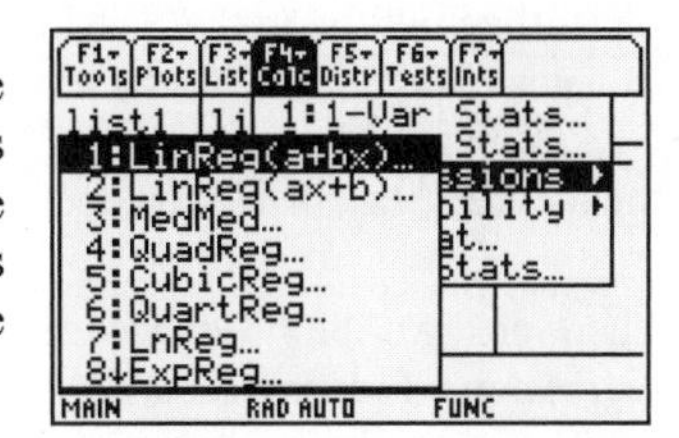

You will be presented with an input screen like those we have seen before. It asks for the list containing the *x* (predictor) variable; the *y* (response) variable; and gives you the option of storing the equation of the line. With the right arrow here you can select none or a *y*-function. Since one usually wants to see the line plotted on the data graph, it is a good idea to select a function (usually `y1(x)`). `Freq` should be left at 1. My regression definition is at right. When finished with the definition, press [ENTER].

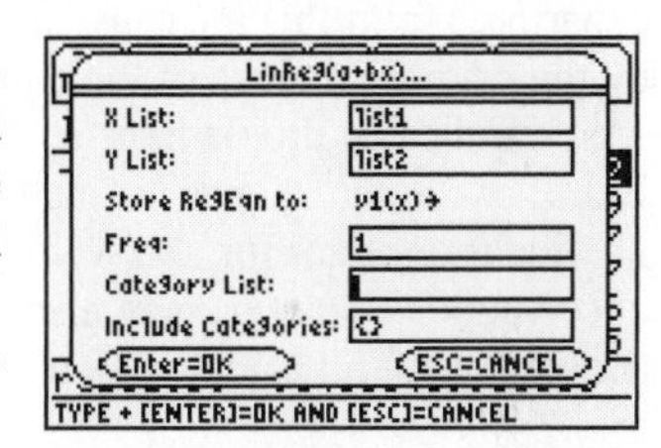

Once the command is executed, you should see the display at right. Notice the first line of the results displays the type of regression in terms of *y* and *x*. Regression lines should never be reported in these terms, but the calculator does not know what variables you are working with. This is really an aid to remind you where the coefficients `a` and `b` go in the equation.

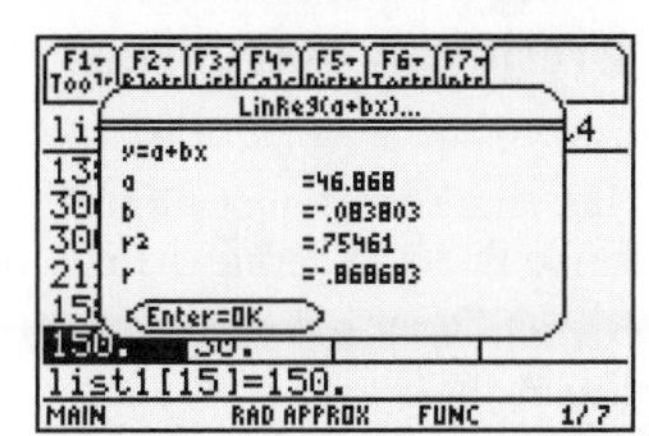

If you press [◆][F1] (`Y=`) you will see the regression equation as on a TI-83/84. Pressing [◆][F3] (`GRAPH`) will graph the data with the line added.

RESIDUALS PLOTS

Residuals are defined to be the vertical distance from the data point to the regression line, in other words, $e_i = y_i - (a + bx_i)$ for each data point (x_i, y_i) in the data set. The e_i are the residuals. There are two ways to obtain the residuals. The first makes use of the stored regression line. From the list of *y*-values we will subtract the value on the line by "plugging in" each corresponding *x*-value into the equation. The residuals will then be stored into a new list.

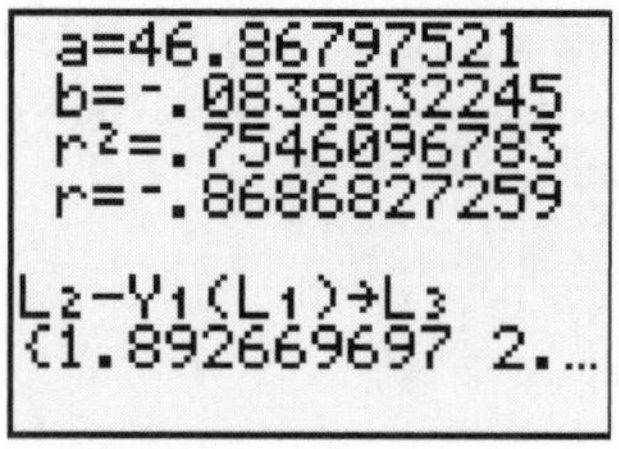

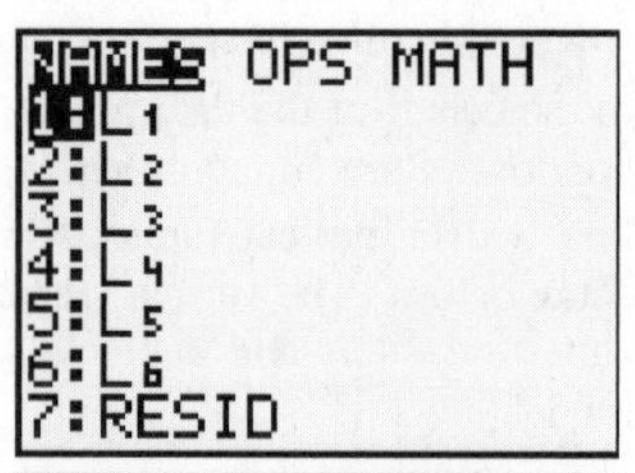

For our example we will enter the following command onto the home screen: [2nd][2] [-] [VARS], arrow to Y-Vars, press [ENTER] to select Function, then [ENTER] to select Y1[(][2nd][1][)][STO▸][2nd][3]. This command says "take the y-list in L2 and from it subtract the value obtained by evaluating function Y_1 at each x value in L1, then store the results into new list L3. Your command should look like the one at right. I have already carried out the command; the first few residuals are displayed. More can be seen by pressing [▸] or by using the STAT editor.

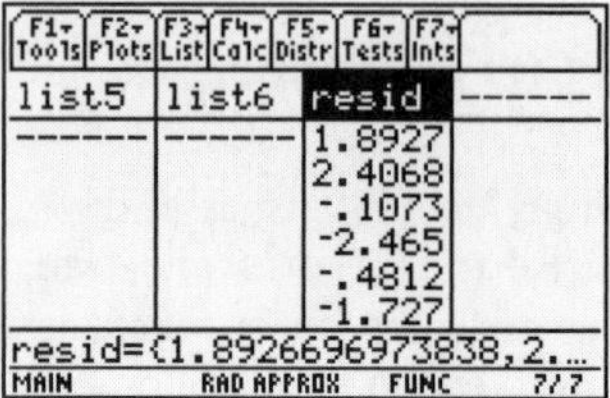

Alternatively, the calculator automatically finds residuals. They can be accessed with [2nd][STAT] (LIST). Notice there is a list called RESID. (You may have many more list names showing on this screen, depending on how the calculator has been used in the past). These are the residuals from the *last* regression performed.

TI-89 calculators automatically add the list of residuals into the Statistics editor when the regression is calculated. When trying to find these in [VAR-LINK] to define plots, they will be in the STATVARS folder. You can use the right and left arrows to expand and hide folder contents. To locate the list more readily, in the STATVARS folder, press [alpha][2]= R to move to the first variable that begins with the letter R (the correlation coefficient).

There are two main types of residuals plots which should be done to examine the adequacy of the model for any regression. The first plots the residuals against x (the predictor variable); the second is a normal probability plot of the residuals. In the first plot, we hope to see random scatter in an even band around the x-axis ($y = 0$ line). Any departures from this are cause for reexamination of the model. In particular, curves may appear which are "masked" by the original scaling of the data; subtraction of any linear trend will magnify any curve. Another common shape which indicates problems is a "fan" in which the plot either narrows from left to right or, alternatively, thickens. Either of the fan shapes means there is a problem with an underlying assumption, namely that the variation around the line is constant for all x-values. If this is the case, a transformation of either y or x is usually necessary. Unusual observations (outliers) may also be seen in these plots as very large positive or negative residuals. Plot definitions are similar for all calculators, with the exception of normal plots on the TI-89 (see the last chapter).

A residuals plot against x (the predictor variable)

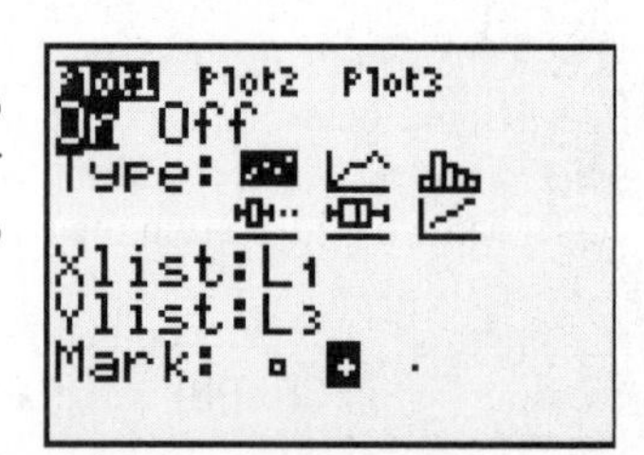

This is a scatter plot. From the plot definitions screen (press [2nd][Y=][ENTER] to define Plot1) define the plot using the original Xlist of the regression (in our example L1) and the residuals list (from our example L3). Press [ZOOM][9] to display the graph.

Alternatively, if using the automatic residuals, define the plot as above, but for the Ylist, press [2nd][STAT] and [ENTER]to select RESID, followed by [ZOOM][9]. The definition screen will look like the one at right.

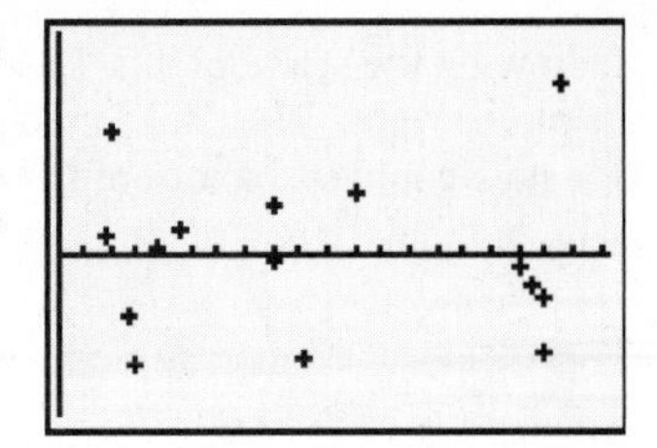

In either case, the residual plot should look like the one at right. Looking at the plot, we see no overt curves, indicating a line appears to be an adequate model. This was a small data set; with these seeing non-constant variation can be difficult. There does not seem to be much of a problem except at the far right end of the graph, where there might be a high outlier, but it's hard to tell.

Normal Probability Plots of Residuals

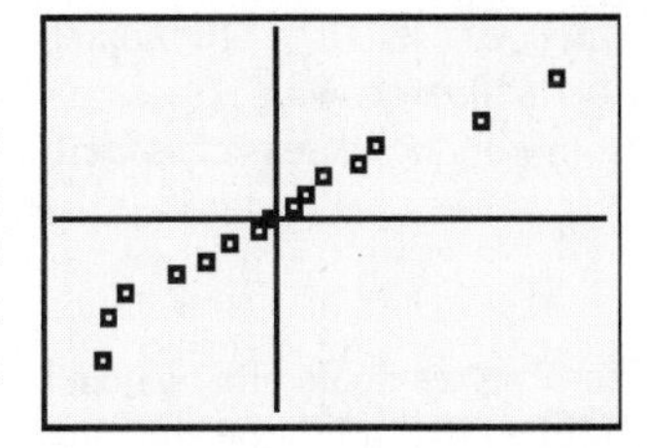

The second plot which should be done is a normal probability plot, since there is an underlying assumption the residuals have a normal distribution. This assumption will be used later in inference for regression. Normal probability plots were discussed in Chapter 4 of this manual. Remember that the data list is the list of residuals. We're looking for an (approximately) straight line. Here, the pattern is fairly linear, indicating no severe problem with this assumption.

Residuals Plots against Time

If the data were gathered through time (the data in our example were not) a time plot of the residuals should be done as discussed in Chapter 2. Ideally, this should look like random scatter. Any obvious patterns (lines, curves, fans, etc) indicate time is an important factor and the model which was fit is not adequate to fully describe the relationship. This generally means a multiple regression is needed to explain the response variable.

USING THE EQUATION TO PREDICT

```
46.87-.08*200
            30.87
Y1(200)
       30.1073303
```

Often we want to predict a value for the response variable based on the regression. This can be done in either of two ways: We can "plug" the value for the predictor (x) variable into the equation explicitly, or with the equation stored as a y-function, simply have the calculator evaluate the result. On the screen at right, we want to know the average gas mileage for a car with a 200-horsepower engine. Notice that since one tends to round the reported slope and intercept, the two answers might disagree. It's much better to have the function evaluate the desired amount using the largest number of significant figures, then round the final result. Based on our linear regression, the average highway gas mileage for cars with 200 horsepower is about 30.1 miles per gallon. We had two cars in our data set that had 200 horsepower: the Audi (32 mpg) and the Buick LaCrosse (30 mpg).

To locate the Y1 function, on a TI-83/84 press [VARS], then arrow to Y-Vars, press [ENTER] to select Function, then [ENTER] again to select Y1. On a TI-89, use [VAR-LINK] to find the variable in the Main folder.

IDENTIFYING INFLUENTIAL OBSERVATIONS

Remember, the large residual on the far right side of the plot looked unusual. Points far away from the center of the range of the predictor variable can be influential; that is, they may have a significant impact on the slope, especially if they do not follow the pattern of the rest of the data. Even if they do not impact the slope, they will cause r and r^2 to be larger than the rest of the data would warrant. To decide if points are influential, delete the suspects, and reanalyze the data.

The large residual corresponded to the Lexus IS 350. It had the highest horsepower rating of all our vehicles, and had a higher highway gas figure than any other vehicle with a similar horsepower rating. Let's delete that data point (be sure to delete both the x and y values from each list) and see if it is influential in some way.

Redrawing the scatterplot (`L1` as `Xlist` and `L2` as `Ylist`) with [ZOOM][9] gives the plot at right. This looks much more linear than the original data scatterplot. One now could suspicious of the cluster of big horsepower vehicles at the lower right, but they're real data!

The new linear regression output is at right. The new regression equation is *Mileage* = 48.97 – 0.10**horsepower*. The original equation was *Mileage* = 46.87 – 0.08**horsepower*. The slope changed by about 20%, which is a fair amount. Both r and r^2 increased as well. The Lexus may well be influential.

```
LinReg
 y=a+bx
 a=48.98613887
 b=-.0962451995
 r²=.8546390943
 r=-.9244669244
```

Here's a residuals plot against horsepower (x). This plot is indicative of another problem that might be encountered in a regression – there is less variability in the residuals as we go from left to right on the graph. We have a violation of the equal variance condition.

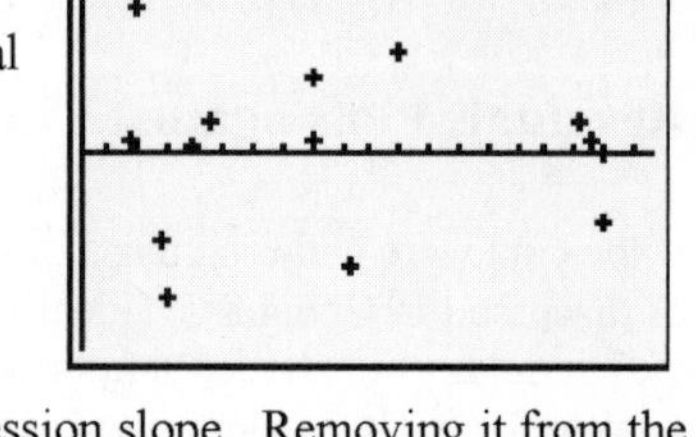

We are left with the following indications: the Lexus was influential on the regression slope. Removing it from the data set impacted the slope (and intercept and r and r^2). However, because of the residual plot above, the indication is that a line is *not* the proper model to describe the relationship between horsepower and gas mileage. What is correct? Perhaps some type of transformation of the data will be better.

TRANSFORMING DATA

There are two reasons to transform data in a regression setting: to straighten a curved relationship and to transform variability so it is constant around the line. In a single variable case, transformations can be used to make skewed distributions look more symmetric; in the case of a single variable observed for several groups, a transformation can make the spread of the different groups look more equal.

The table below shows stopping distances in feet for a car tested three times at each of five speeds. We hope to create a model that predicts stopping distance from the speed of the car. (Data are from problem 17 in Chapter 10 of the text.)

Speed (mph)	Stopping Distance (ft)
20	64, 62, 59
30	114, 118, 105
40	153, 171, 165
50	231, 203, 238
60	317, 321, 276

A plot of the data is at right. It looks fairly linear, but it is clear that the stopping distances become more variable with faster speed.

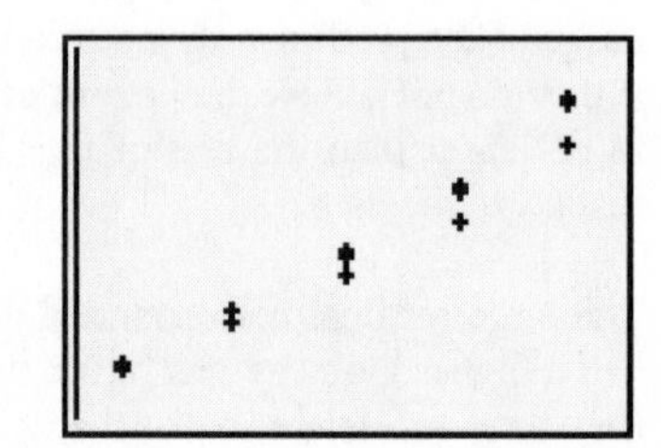

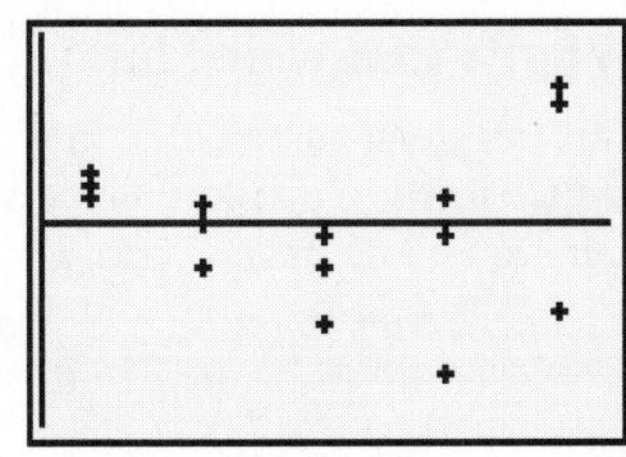

Regression gives the fitted model as *Stoppingfeet* = −65.933+5.98**Speed*. The residuals plot against speed (at right) clearly indicates the variability gets larger for faster speeds; it also indicates the true relationship is not linear but curved. Clearly, a transform is indicated – one which will decrease variation as well as straighten the plot.

Since the residuals indicate a curve (possibly quadratic), using the square root of stopping distance makes sense. With stopping distance in L2, we need to find the square root of each distance. With one command we can do this, storing the result in a new list, say L3. Press [2nd][x^2] ([√]) [2nd][2][)] [STO▸][2nd][3] followed by [ENTER]. The command and result are at right. We see the first couple values. How many are shown depends on the number of decimal places displayed. To see the entire list, go to the STAT editor.

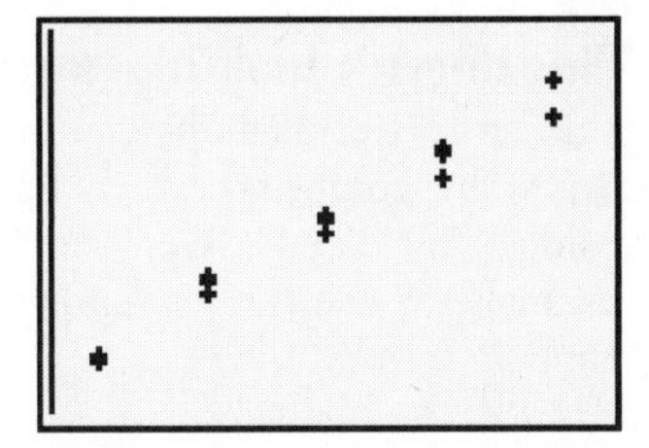

The new scatterplot is at right. The new regression equation is *sqrt*(*Stoppingfeet*) = 3.303 + 0.235**Speed* We have $r = 0.9922$, an extremely strong linear relationship. What about a residuals plot?

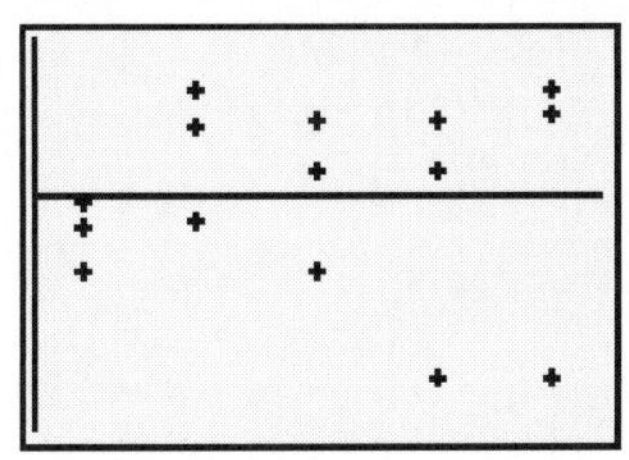

Here is the residuals plot. It's not perfect; the variation still increases with larger values of speed, but is much better than before. Sometimes there is no "perfect" transform.

WHAT CAN GO WRONG?

What's Dim Mismatch?

We've seen this one before. Press [ENTER] to quit. This error means the two lists referenced (either in a plot or a regression command) are not the same length. Go to the STAT editor and fix the problem.

What is Err: Invalid?

This error is caused by referencing the function for the line when it has not been stored. Recalculate the regression being sure to store the equation into a y function.

What's that weird line?

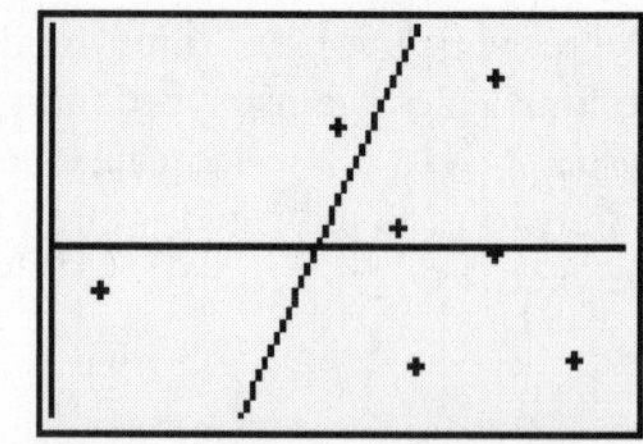

This error can come either in a data plot (an old line still resides in the [Y=] screen) or the stored regression line is showing in the residuals plot, as shown here. The regression line is not part of the residuals plot and shows only because the calculator tries to graph everything it knows about. Press [Y=] followed by [CLEAR] to erase the unwanted equations, then redraw the graph by pressing [GRAPH].

What does Nonreal Ans mean?

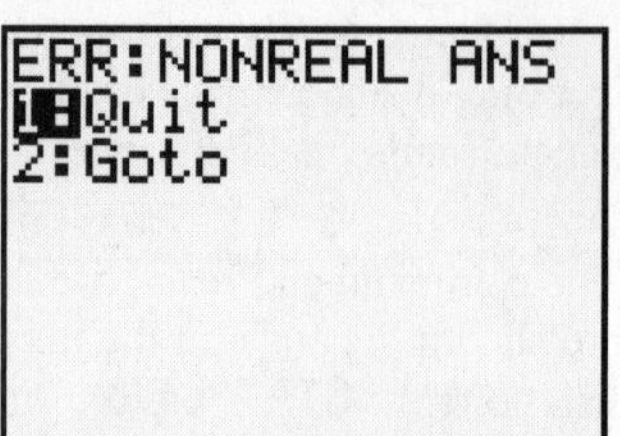

This error comes from trying to take the square root (or log) of a negative number. This can't be done in the real number system. These transforms do not work for negative values. Try something else.

This doesn't look like a residuals plot!

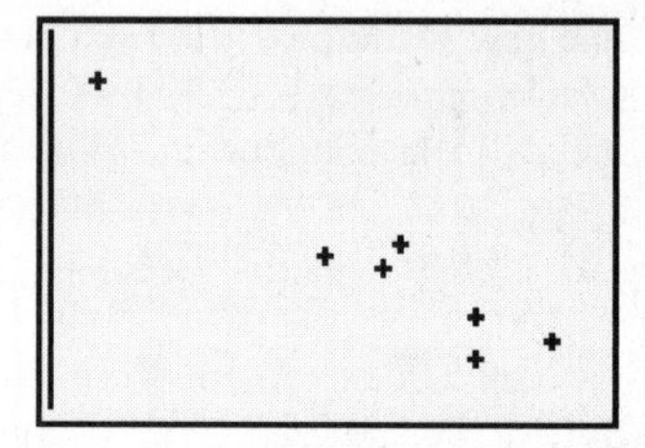

It doesn't. Residuals plots *must* be centered around $y = 0$. This error is usually caused by confusing which list contains the y's and which the x's in finding residuals by "hand." Go back and check which list is which, then recomputed the residuals, or use the list automatically stored by the calculator (under the LIST menu.

Chapter 6 – Random Numbers

Randomness is something most people seem to have an intuitive sense about. But truly random values are surprisingly hard to get. In fact, calculators (and computers) can't generate true random numbers since any values they obtain are based on an algorithm (that is, a program). But they do give good *pseudorandom* numbers. These have many applications from simulation to selecting samples and assigning treatments in an experiment. Random selection (and random number generation) form the basis of sample selection (choosing randomly from the population) and treatment assignment in experiments (to avoid bias.)

SIMULATIONS

Simulations are used to mimic a real situation such as this. Suppose a cereal manufacturer puts pictures of famous athletes in boxes of their cereal as a marketing ploy. They announce that 20% of the boxes contain a picture of Tiger Woods, 30% a picture of David Beckham, and the rest have a picture of Serena Williams. Assuming the pictures are distributed in the boxes at random in the specified ratios, how many boxes of the cereal do you expect to have to buy in order to get a complete set?

You could go out and buy lots of cereal, but that might be expensive. We'll model the situation using random numbers, assuming the pictures really are randomly placed in the cereal boxes, and distributed randomly to stores across the country.

We'll use random digits to represent getting the pictures: since 20% have Tiger's picture, we'll let the digits 0 and 1 represent getting his picture. Similarly, we'll use digits 2, 3, and 4 (30% of the ten digits) to represent getting Beckham's picture. The rest (5 through 9) will mean we got a picture of Serena. We need to get the random digits.

TI-83/84 Procedure

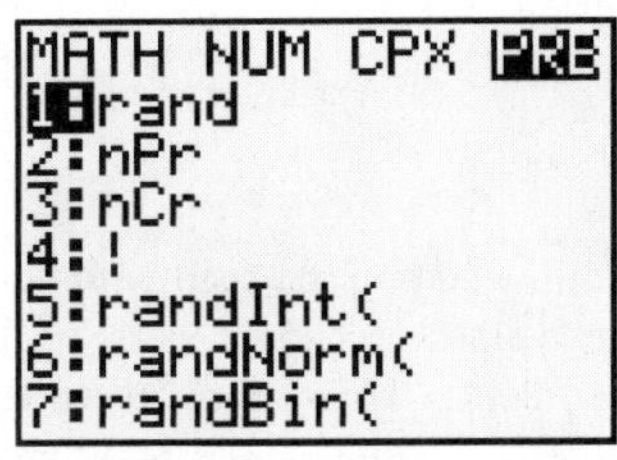

From the home screen, press [MATH], then arrow to PRB. The menu at right is displayed. We want choice `5:randInt(`. Either arrow to it and press [ENTER] or simply press [5]. The command shell will be transferred to the home screen. Now you need to tell the calculator the boundary values you want. Since we want digits between 0 and 9, we will enter [0][,][9] . Pressing [ENTER] will get the first random digit.

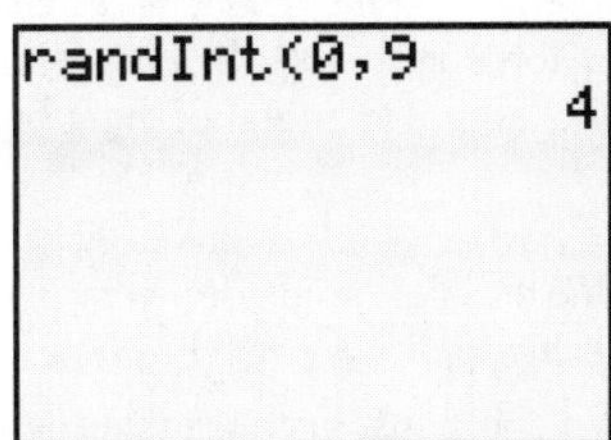

Here is our first random digit: a 4. That means the first box had a picture of Beckham. We can continue pressing [ENTER] and get more random digits.

TI-89 Procedure

From the home screen, press [2nd][5] (MATH), then arrow to `7:Probability`. Pressing the right arrow displays the menu at right. We want choice `4:rand(`. Either arrow to it and press [ENTER] or simply press [4]. The command shell will be transferred to the home screen. This command generates numbers between 1 and the value specified, so here we would renumber our possibilities and consider a 1 or 2 as getting Tiger's picture; 3, 4, or 5 Beckham's picture, and values 6 through 10 as Serena's picture. Since we want numbers between 1 and 10, we will enter [1][0][)] . Pressing [ENTER] will get the first random number.

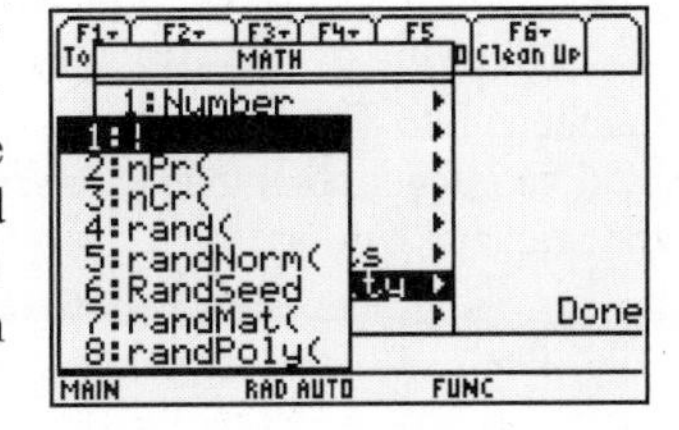

Here is our first random number: a 10. That means the first box had a picture of Serena. We can continue pressing [ENTER] and get more random digits.

In general, for random integers between 1 and *N*, the parameter for the rand command is *N*.

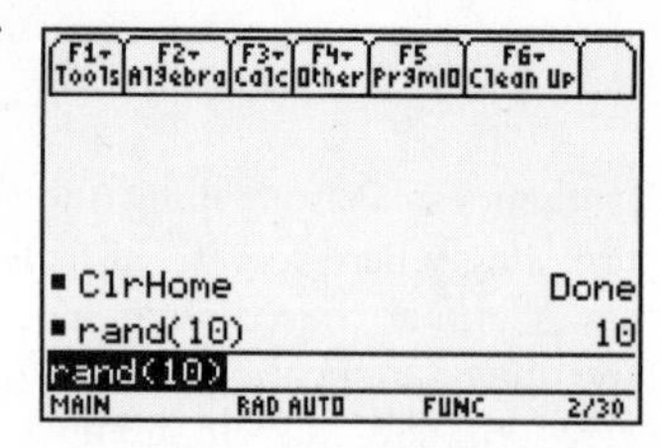

CONTROLLING THE SEQUENCE OF RANDOM DIGITS

You didn't get the same random number? Not surprising. Random number generation on computers and calculators works from something called a *seed*. In the case of TI calculators, every command you use changes the seed. If a value is explicitly stored as the seed *immediately before* a random number command, the sequence of random digits will be the same every time.

To store a seed, enter the value desired, then press [STO▸][MATH], then arrow to **PRB** and press [ENTER] to select rand then press [ENTER] again to actually store the seed. The sample at right stored 1187 as the seed.

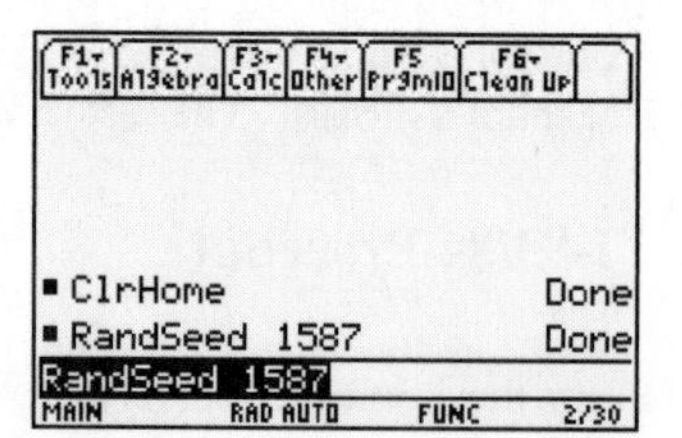

Good seeds are large, preferably prime (or at least odd) numbers.

To store a seed on a TI-89, go to the Math, Probability menu as before, but select choice `6:RandSeed.` Type in the desired seed value and press [ENTER]. Here, the seed value is 1587. Press [CLEAR] to erase the command from the entry area.

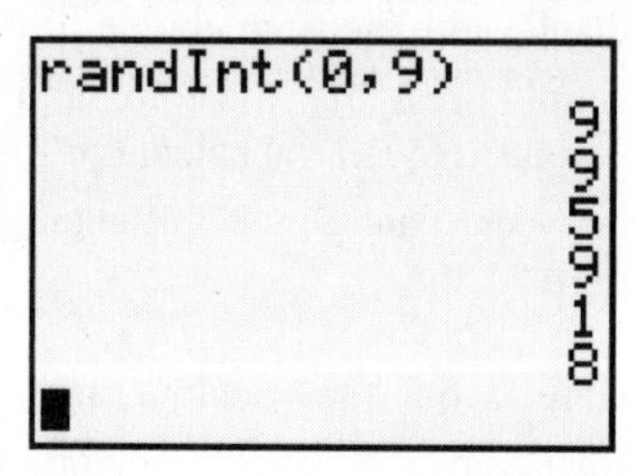

Follow setting the seed with the same random integer command as before, and we get values as shown at right. Yours should be the same. Look at the first five numbers. These correspond to (in our example above) getting Serena, Serena, Serena, Serena, Tiger, and Serena. Even after the sixth "box" we haven't gotten all three pictures. In fact, it takes two more boxes (another Tiger then finally a Beckham) for a total of eight boxes to get the complete set.

One simulation is not a very good representation. We'd like to know how many boxes it would take to get all three, *on average*. We need to repeat the simulation many times, and take the average value from the many simulations. We could just keep pressing [ENTER] until we've done enough, or we can get many random numbers at once and store them into a list.

Here, we've reset the seed to 18763 and changed the random integer command to add another parameter – how many numbers to generate. Since it could possibly take many boxes of cereal to get all three pictures, we've chosen to store 200 numbers into list `L1`. (Press [STO▸][2nd][1] after the ending parenthesis on the random integer command.) The first few values are displayed. To see the rest, use the Statistics Editor.

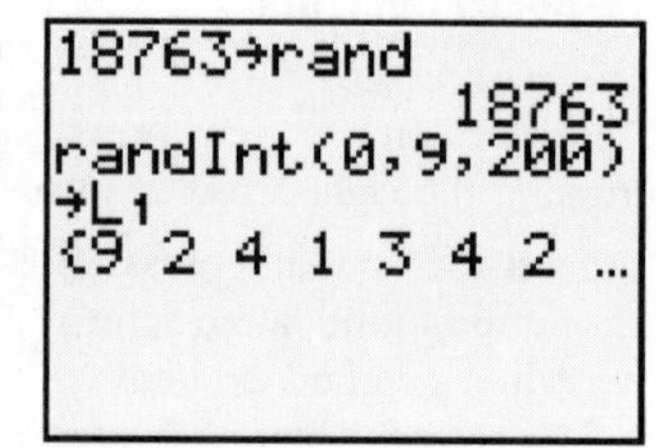

On a TI-89, inside the Statistics Editor, move the cursor to highlight the name of a list. Press [F4] then [4] for Probability and select choice `5:randInt(`. The parameters are the low number to be generated, the high end of the desired numbers, and how many. End the command by closing the parentheses. Since it could possibly take many boxes of cereal to get all three pictures, we've chosen to store 200 numbers into list1. Pressing [ENTER] executes the command and fills in `list1`. To set the seed here, select menu option `A:RandSeed` from the Probability menu.

Looking at the list in the editor, the first four digits are 9, 2, 4, and 1. That corresponds to a Serena, Beckham, Beckham, and Tiger. That trial took four boxes to get the full set. One can continue down the list until several complete sets have been found; then compute the average for all trials as the estimate of the average number of boxes.

RANDOM NORMAL DATA

These calculators can also simulate observations from normal populations in a manner similar to the examples above. The command is choice `6:randNorm(` from the `Math`, `Prb` menu. The parameters are the mean and standard deviation.

This example models the following: A tire manufacturer believes that the tread life of their snow tires can be described by a Normal model with mean 32000 miles and standard deviation 2500 miles. You buy 4 of these tires, hoping to drive them at least 30000 miles. Estimate the chances that all four last at least that long. We have output for one trial – a set of four tires. In this trial, 3 of the 4 lasted over 30000 miles. To further estimate the chance that all four last over 30000 miles, obtain more repetitions of sets of four tires.

With a TI-89, random normal data are option 6 on the Probability menu in the Statistics/List Editor application. Specify the mean, standard deviation, and the number of values you want similarly to the random integer command.

SAMPLING AND TREATMENT ASSIGNMENTS

Random numbers are the best method for (randomly!) selecting items or individuals to be sampled or treatments to be assigned in an experiment. In the sampling frame (a list of members of the population) number the individuals from 1 to N, where N is the total number in the list. Use the Random integer command to select those to be sampled. In the case of assigning treatments, if there are, for example two treatments, use random integers to assign half the experimental units to treatment A; the rest will get treatment B.

THE PROB SIM APPLICATION

TI-83+ Silver Edition calculators and the TI-84 series also come with the Prob Sim application. Press [APPS]. From the list of applications, select Prob Sim. Press [ENTER] to select it, then any key to get the simulation main menu.

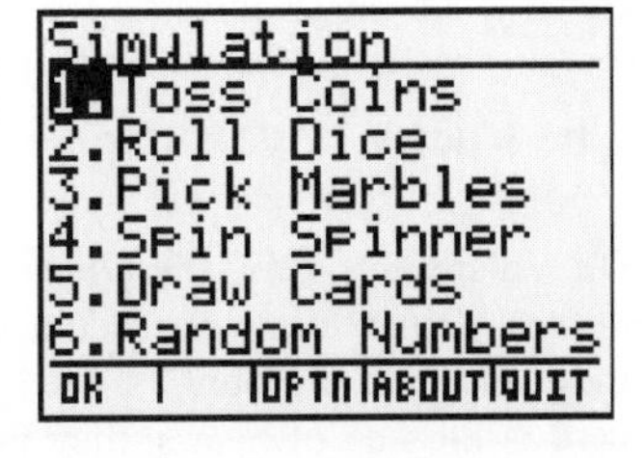

We can use menu option 6 to mimic the room lottery drawing in the text. Press [6] then [ZOOM](`Set`) to set the parameters for the simulation. There are a total of 57 students, of which we will consider numbers between 1 and 20 to represent the varsity athletes. Any number 21 or larger will be a regular student. On the set-up screen, tell the simulator you want 3 numbers at a time between 1 and 57 with no repeats, then press [GRAPH] for `OK`. Press [WINDOW] (`Draw`) repeatedly to draw the room lottery numbers. Follow the program prompts to exit the simulator, change parameters, or change simulations. Unfortunately, this application turns off screen capture, so I can't show you my results.

Chapter 7 – Random Variables

TI Calculators can help find the mean and standard deviation for random variables, given a probability distribution. This basically uses `1-Var Stats` as described in Chapter 2 using the frequencies as probabilities.

Suppose, for example, that the death rate in any year is 1 out of every 1000 people, and that another 2 out of 1000 suffer some kind of disability. An insurance company will pay \$10,000 for a death and \$5000 for a disability. To see what the company can expect to pay, we will first enter the payouts in L1 and the probabilities in L2.

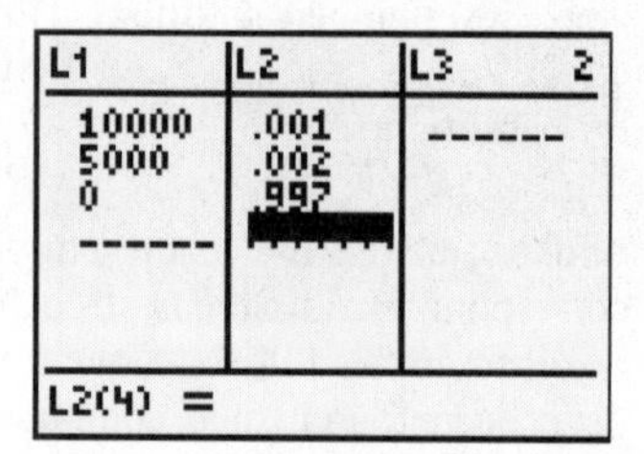

Note that nothing should happen 997 times in 1000, and the company has to pay nothing.

From the STAT, CALC menu, select option `1:1Var Stats`. Specify the two lists: first the list of the values, and then the list of the probabilities.

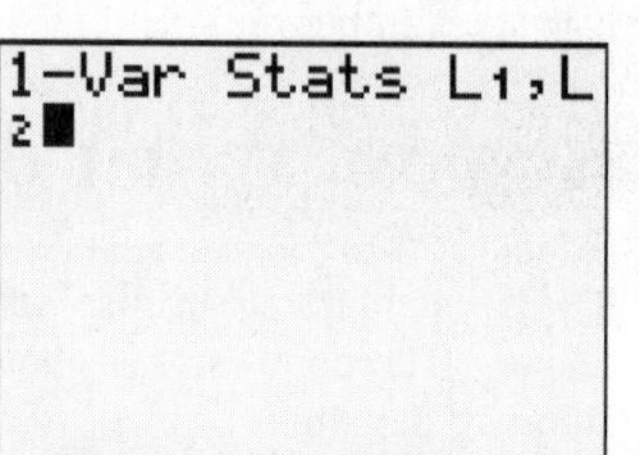

If you are using a TI-89, from the [F4] `Calc` menu, select `1:1-Var Stats`. In the input boxes, I have specified that the values are in `list1` and the probabilities (frequencies) are in `list2`.

```
1-Var Stats...
List:              list1
Freq:              list2
Category List:
Include Categories: {}
Enter=OK           ESC=CANCEL
list2[4]=
MAIN   RAD APPROX   FUNC   2/8
```

The results show that the company can expect to have to pay \$20 per policy, on average. The standard deviation of this value is \$386.78. Why the large standard deviation? Consider the spread from a payout of \$0 to \$10,000 – that's a lot!

```
1-Var Stats
 x̄=20
 Σx=20
 Σx²=150000
 Sx=
 σx=386.7815921
↓n=1
```

Notice that the calculator doesn't give a sample standard deviation – working with decimal (or fractional) frequencies, it understands we are looking at relative frequencies (probabilities).

If you want the variance, square the standard deviation. Here, we would find that $(386.78)^2 = 149598.7684$. The text gives the variance as 149600 – the minor difference is due to my rounding the calculator's standard deviation before squaring – if in doubt, as always, use all the digits for intermediate calculations, then round at the end!

Be sure that the total frequency (n) is 1. If not, you have made an error entering your probabilities.

The Lucky Lover's Special

On Valentine's Day, the *Quiet Nook* Restaurant offers a *Lucky Lover's Special* that could save couples on their dinner. Diners draw at random from a deck containing only four aces. If the first card drawn in the ace of hearts, they get a \$20 discount. If the first card drawn is the ace of diamonds, they will get another chance. If the second card is the ace of hearts, they get a \$10 discount. If their first card drawn is one of the black aces, they get no discount. What might couples expect the average discount to be, and how variable is it?

The hardest part of this problem will be finding the probabilities for most people. Clearly, since we are only dealing with the four aces, P(Ace of Hearts) = 1/4 which gets the \$20 discount. Getting the ace of diamonds on the first card also has probability 1/4; if this card is drawn, you get another chance to draw. At this point, there are 3 cards left, so the probability of following with the Ace of hearts is 1/3, but we need to multiply the two probabilities for the sequence of events. Therefore, P(Ace of diamonds followed by Ace of hearts) = (1/4)(1/3) = 1/12 which gets a $10 discount. Lastly, anything else gets no discount. Recognizing that getting no discount is the complement of either of the two preceding situations, we have P(No Discount) = 1 – (1/4 + 1/12) = 2/3.

Here, I've entered the possible discounts and the probabilities into lists L1 and L2. Because of the repeating decimals in 1/12 and 2/3, I entered them as the fractions, letting the calculator convert to fractions. What displays is rounded, but the calculator will really use the fractional equivalent.

L1	L2	L3 2
20	.25	------
10	.08333	
0	.66667	

L2(4) =

Here are the results. The typical (average) discount will be about $5.83. The standard deviation of the discounts is $8.62. The standard deviation is still larger than the average as in the preceding example, but not so dramatic as before, because the range of possible discounts is much smaller.

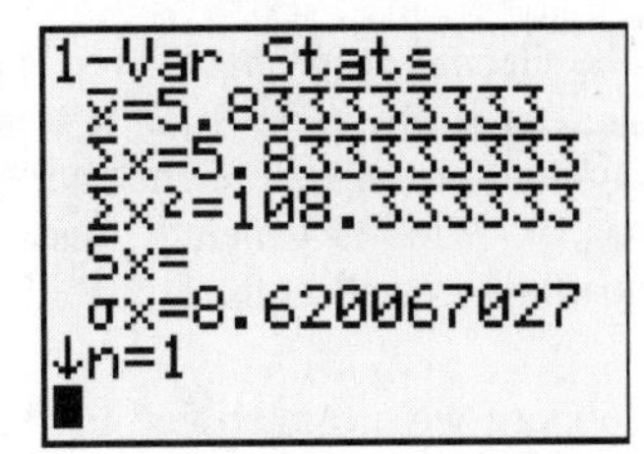

WHAT CAN GO WRONG?

My answer's not the same!

Errors when dealing with distributions like this usually come from forgetting to specify the second list of probabilities. Notice that on this screen we have n=3, not n=1. The correct form of the command is `1-Var Stats xlist,plist`, where `xlist` is where the actual values were stored and `plist` is where the frequencies are.

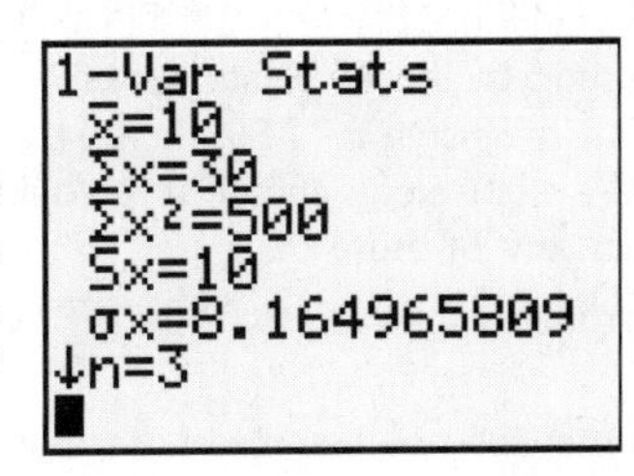

Chapter 8 – Probability Models

We've already used the calculator to find probabilities based on normal models. There are many more models which are useful. This chapter explores three such models.

Many types of random variables are based on Bernoulli trials experiments. These involve independent trials with only two outcomes possible, and a constant probability of success called p. Two of the more common of these variables have either a Geometric or Binomial model. The Poisson random variable is related as well – it can be viewed as a Binomial model with small probability of success and large n, or as a process where only "successes" are seen: phone call arrivals at a switchboard, for example.

GEOMETRIC MODELS

The Geometric probability model is used to find the chance the first success occurs on the nth trial. If the first success is on the nth trial, it was preceded by n - 1 failures. Because trials are independent we can multiply the probabilities of failure and success on each trial so $P(X = n) = (1-p)^{n-1}p$ which is sometimes written as $P(X = n) = (q)^{n-1}p$. This is generally easy enough to find explicitly, but the calculator has a built-in function to find this quantity, as well as the probability of the first success occurring somewhere on or before the nth trial.

Suppose we are interested in finding blood donors with O-negative blood; these are called "universal donors." Only about 6% of people have O-negative blood. In testing a group of people, what is the probability the first O-negative person is found on the 4th person tested? We want $P(X = 4)$. Press [2nd][VARS] (`Distr`). If you are using a TI-89, the `Distr` menu is [F5] in the Stats/List Editor application. We want menu choice `D:geometpdf (F` on an 89). To find this menu option (or any of those discussed in this chapter), pressing the up arrow to find them is probably the easiest. Press [ENTER] to select the option.

Enter the two parameters for the command: p and x (n). Here, p = .06 and n = 4. Press [ENTER] to find the result. We see there is about a 5% chance to find the first O-negative person on the 4th person tested (assuming of course that the individuals being tested are independent of each other).

There are some other related questions that can be asked. What is the probability the first O-negative person will be found in the first four persons tested? We want to know $P(X \leq 4)$. We could find all the individual probabilities for 1, 2, 3, and 4, and add them together but there is an easier way. We really want to "accumulate" all those probabilities into one, or find the *cumulative* probability.

This uses menu choice `E:Geometcdf (G` on an 89). The cdf part stands for cumulative distribution function. On a TI-83/84 we input the probability of a success (still .06); and the high end of interest (4). Using a TI-89, you will be prompted for the low end of interest (`Lower Value`) and the upper end of interest (`Upper Value`).

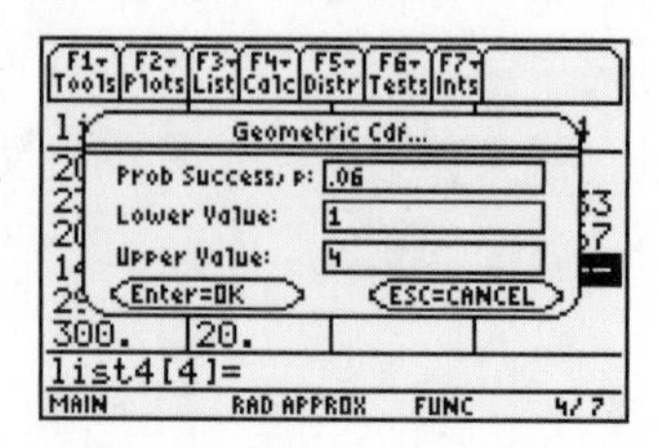

After pressing [ENTER] we see that finding the first O-negative person within the first 4 people tested should happen about 21.9% of the time.

```
1-geometcdf(.06,
9)
          .5729948022
```

What's the chance we'll have to test at least ten before we find one with type O-negative blood? We want $P(X \geq 10)$. Since there could (possibly) be an infinite number of people tested to find the first O-negative person. Using a TI-83/84 we will use complements to answer this question. The complement (opposite) of needing at least ten people to find the first universal donor is finding the first one somewhere in the first nine tested. Using the complements rule, $P(X \geq 10) = 1 - P(X \leq 9)$. It makes no difference whether we find $P(X \leq 9)$ and then subtract from 1 or do both operations in a single step.

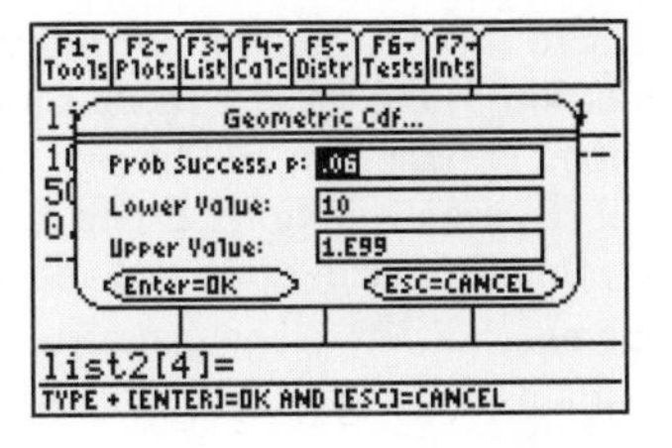

Using a TI-89, since the lower end of interest and the upper end of interest are explicitly asked for, here we will say the lower end is 10, and the upper end is [1][EE][9][9] for infinity.

The probability we will find the first O-negative person somewhere on the tenth person or later is 57.3%.

Spam and Geometric Models

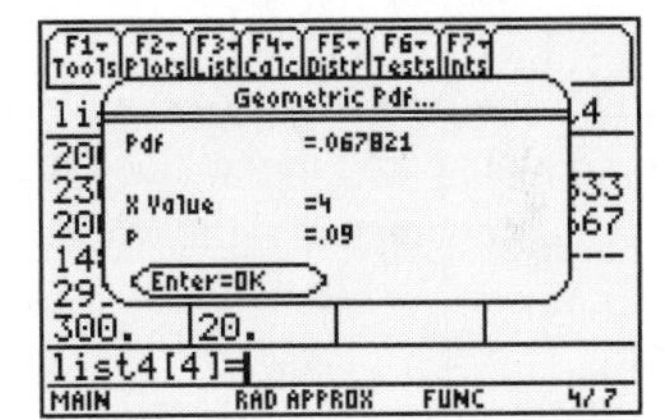

According to *Postini*, about 91% of all email traffic is spam. That means only about 9% is legitimate email. Since spam comes from so many sources, assuming messages are independent is reasonable. What's the chance you have to check four messages in your inbox to find the first legitimate message? According to the calculation at right, abut 6.8%.

BINOMIAL MODELS

Binomial models are interested in the chance of k successes occurring when there are a fixed number (n) of Bernoulli trials.

Suppose you plan on buying five boxes of cereal that each have a 20% chance of having a picture of Tiger Woods in them. Assuming the boxes are chosen at random, they should be independent of one another. We are interested in the probability that there are exactly two pictures of Tiger in the five boxes. How many arrangements are possible so that there are two Tiger pictures? You could get the two on the first two boxes and not on the last three, or on the second and fifth, or on the third and fourth, etc. How many arrangements are possible to get exactly two pictures of Tiger and three of someone else? The Binomial coefficient provides the answer to this question. The coefficient itself is variously written as $\binom{n}{r}$ or nCr and is read as "n choose r."

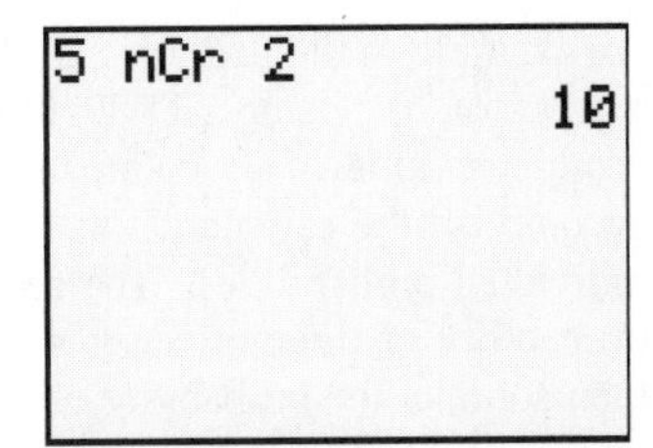

On the TI-83/84 home screen, type in the desired n (5, here), then press [MATH] then arrow to **PRB**. Either arrow down to **3:nCr(** and press [ENTER] or just press [3]. The command is transferred to the home screen. Finally, enter k (3 here). Execute the command by pressing [ENTER].

```
rand(10)          10.
rand(10)           2.
rand(10)           6.
rand(10)           5.
rand(10)           8.
nCr(5,2)          10.
nCr(5,2)
MAIN   RAD APPROX   FUNC   30/30
```

If you are using a TI-89, there are two ways to find the number of combinations. On the home screen, press [2nd][5] for the **Math** menu, then press [4] to expand the probability menu. Then choose **nCr**, which is menu option 3. Press [3] to transfer the shell to the input area. Type in n and r separated by a comma and close the parentheses. Press [ENTER] to complete the computation. You can also perform the calculation in the in the same manner while in the Statistics/List

Editor using the `Probability` menu under [F4] (`Calc`).

We find there are 10 possible ways to get exactly two pictures of Tiger in five boxes of cereal. We can then find the probability of exactly two pictures of Tiger as $10(.2)^2(.8)^3 = 0.2048$. This means about 20% of the time, if you bought five boxes of cereal (independently of each other) you would get two pictures of Tiger.

```
5 nCr 2
                10
10*.2²*.8^3
             .2048
```

There is an easier way to find these binomial probabilities. Looking back at the portion of the `Distr` menu shown on the first page of this chapter, there are two menu options that will help us here: `Ø:Binompdf` and `A:Binomcdf`. On a TI-89, these are options `B` and `C` in the `Distr` menu. Just as in the case of the geometric model discussed above, the pdf menu choice gives $P(X = x)$ and the cdf gives $P(X \le x)$. The parameters for both commands are n, p, x. Three examples follow.

```
DISTR DRAW
9↑Fcdf(
Ø:binompdf(
A:binomcdf(
B:poissonpdf(
C:poissoncdf(
D:geometpdf(
E:geometcdf(
```

These commands are similar on a TI-89, with the exception that for the cdf one explicitly specifies the lower value of interest and the upper value of interest, just as we did with the geometric model.

Returning to the prior example about blood donors, what is the probability that if 20 donors come to the blood drive, there will be exactly three O-negative donors? From the screen at right, we see this is 8.6%.

```
binompdf(20,.06,
3)
       .0860066662
```

As with geometric models, we can find the use the cdf command to find cumulative probabilities. Remember, the calculator adds individual terms for the values of interest specified. For instance, what is the chance of at most three O-negatives in a blood drive with 20 donors? We see this is very likely to happen. We'll expect three 3 or fewer O-negative donors in 20 people about 97% of the time.

```
binomcdf(20,.06,
3)
       .9710342619
```

What's the chance there would be more than two O-negative donors in a group of 20? Again, since we are looking for $P(X > 2)$ we use complements. The complement of more than 2 is 2 or less. So we find $P(X > 2)$ as $1 - P(X \le 2)$.

```
1-binomcdf(20,.0
6,2)
       .1149724038
```

An example in the text asks for the probability of two or three universal donors in a group of 20. Using a TI-89, this could be done specifying a lower value of 2 and an upper value of 3. With the TI83/84 series, there are two possible ways to accomplish the calculation as shown at right. The first explicitly adds the two individual probabilities. The second uses the binomial cdf function and finds the probability of three or fewer which includes the possibility of 0, 1, 2, or 3. We then subtract the probability of 0 or 1 from the result because we are interested in only 2 or 3. This method of calculation is very efficient in case you wanted to find, for example, the probability of between 12 and 19 universal donors – adding individual terms would become tedious. Find the cumulative probability for the upper end of interest, then subtract what was included that is not of interest.

```
binompdf(20,.06,
2)+binompdf(20,.
06,3)
       .3105796278
binomcdf(20,.06,
3)-binomcdf(20,.
06,1)
```

Binomial Distributions with large n.

Older TI-83 calculators (and most computer applications as well) cannot deal with large values of n. This is because the binomial coefficient becomes too large very quickly. However, when n and p are sufficiently large (generally, *both* of $np \geq 10$ and $n(1-p) \geq 10$ must be true to move the distribution away from the ends so it can become symmetric) binomials can be approximated with a normal model. One uses the `normalcdf` command described in Chapter 4 specifying the mean as the mean of the binomial ($\mu = np$) and the standard deviation as that of the binomial ($\sigma = \sqrt{np(1-p)}$). Newer models of the TI-83 (and 84) actually use an approximation to find the probabilities in these situations.

Suppose the Red Cross anticipates the need for at least 1850 units of O-negative blood this year. They anticipate having about 32,000 donors. What is the chance they will not get enough? We desire P(X < 1850). We have calculated the mean to be 1920 and the standard deviation to be 42.483. On this scale, 1850 has a z-score of -1.648. It appears there is about a 5% chance there will not be enough O-negative in the scenario discussed.

Spam and Binomial Models

```
binompdf(25,.09,
1)+binompdf(25,.
09,2)
         .5116672473
```

Suppose there were 25 new email messages in your inbox. As we've seen before, there's only about a 9% chance that any one of them is real (assuming you're not using a good filter). What's the chance you'll find only one or two good messages? This calculation is similar to the one done above where we were interested in 2 or 3 universal blood donors. We need the sum of two binomial results. From the screen at right, we see the probability of only 1 or 2 real messages in 25 is about 51.2%.

POISSON MODELS

Poisson models can be viewed in terms of several scenarios: an approximation to the binomial when the number of trials is large and the probability of a success is small (like the Woburn leukemia example in the text) or as realizations of a process in which we can really only see the successes. Classic examples are phone calls to a switchboard (we know about those who actually got through but not people who got busy signals, thought about calling but didn't, etc) and catching fish in a lake (we don't know about the fish who maybe ignored the bait). Lucky for us that the name of the individual who developed the model was Simeon Poisson (his name is French for fish).

Poisson models are generally characterized by a rate called λ (lambda). In the binomial setting, this is the mean of the binomial variable, $\lambda = \mu = np$. In the other setting, we think in terms of the rate of occurrence: we normally get 10 phone calls per hour, or I can typically catch 5 fish per day.

```
.00011*35000
                3.85
poissoncdf(3.85,
7)
         .9572996811
1-Ans
         .0427003189
```

Revisiting the Woburn example, the national incidence rate for new leukemia cases each year is about 0.00011. That's p. The town of Woburn had about 35,000 people. Multiplying the two tells us that we'd expect about 3.85 new leukemia cases per year in a town that size. We want to find the probability of at least 8 cases in a single year. Since 8 or more is the complement of 7 or less, we use the calculator and first find the probability of 7 or less cases (using `poissoncdf` from the `DISTR` menu) , then subtract that result from 1. The chance of at least 8 new cases is about 4.3%. That result is pretty small (less than 1 in 20 chance), but not totally unreasonable.

WHAT CAN GO WRONG?

What does Err: Domain mean?

This error is normally caused in these types of problems by specifying a probability as a number greater than 1 (in percent possibly instead of a decimal) or a value for n or x which is not an integer. Reenter the command giving p in decimal form. Pressing [ESC] will return you to the input screen to correct the error. This will also occur in older TI-83 calculators if n is too large in a binomial calculation; in that case, you need to use the normal approximation.

ERR:DOMAIN
1:Quit
2:Goto

How can the probability be more than 1?

It can't. As we've said before, if it looks more than 1 on the first glance, check the right hand side. This value is 9.7×10^{-18} or seventeen zeros followed by the leading 9.

binomcdf(1800,.25,300)
9.69848117E-18

Chapter 9 – Sampling Distribution Models

We know the sample proportion of observed "successes," $\hat{p}$, is a random variable. We don't know what value we'll find until we actually take a sample. Likewise, we also can recognize that if a different sample were taken (involving different individuals or objects) we'd most likely get a different value. We need a way to model the distributions of sample statistics like $\hat{p}$ or the sample mean $\bar{x}$. This will form the basis for statistical inference to follow.

Intuitively, the mean for a statistic such as $\hat{p}$ (that estimates the population proportion p) should be close to p if we have a good (random and independent) sample. Similarly, $\bar{x}$ should be close to the population mean μ. How close? That's measured by the standard deviation. The standard deviation of $\hat{p}$ is defined (actually mathematically proven) to be $\sigma(\hat{p}) = \sqrt{\frac{p(1-p)}{n}} = \sqrt{\frac{pq}{n}}$. It can be shown through simulation that if both $np \geq 10$ and $n(1-p) \geq 10$ that $\hat{p}$ will be approximately normally distributed. Similarly, we find (for large samples, or in the case of a normal population) that $\bar{x}$ has a normal distribution with mean μ and standard deviation $\sigma(\bar{x}) = \frac{\sigma}{\sqrt{n}}$.

How can we make use of these facts? Ultimately for inference – making conclusions about what we believe is true about the actual population parameters. For now, we'll use these ideas for some probability questions.

USING SAMPLING DISTRIBUTIONS FOR A PROPORTION

The Centers for Disease Control report that 22% of 18-year-old women have a body mass index (BMI) of 25 or more. These values are associated with increased health risk. As part of a routine health check at a large college, the physical education department decided to try a self-report system, instead of having girls actually come in to be measured (height and weight since BMI = weight in kg/(height in m)^2, they asked 200 randomly selected female students to report their heights and weights. Only 31 had a BMI higher than 25. Should the physical education department continue to use the self-reporting system?

The sample fulfills the conditions necessary for doing the calculations needed. We had 31 females with BMI over 25, so there must have been 169 with BMI 25 or less (according to their reported heights and weights.) The females were randomly selected, and it's reasonable that they are less than 10% of all females at this "large college." First of all, the observed proportion of the randomly selected females is $\hat{p}$ = 31/200 = 15.5%. This seems low. What's the chance of getting this type of result (or something more extreme) if the national rate really holds at this university? If the national rate is true here, we should have $p = 0.22$. This is the mean for the normal distribution. The standard deviation will be $\sigma(\hat{p}) = \sqrt{\frac{.22(.78)}{200}} = 0.02929$.

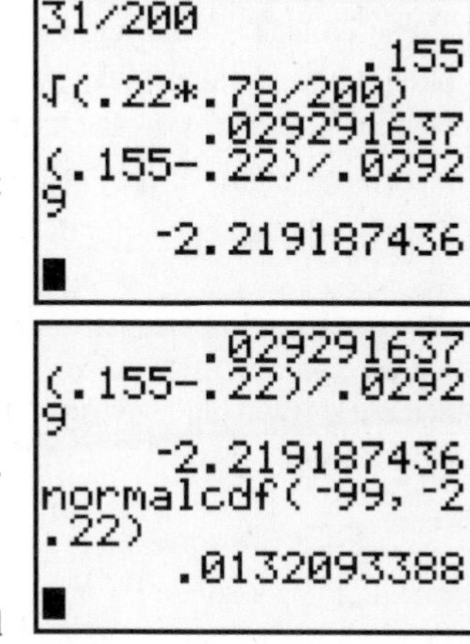

To find the chance of our results (assuming, of course that the national model holds) we'll first compute the z-score for our observed result. Our observed 15.5% is about 2.2 standard deviations *below* the expected 22%. Is this unusual? Based on the 68-95-99.7 Rule, we can say it is. How unusual is it, exactly?

We're back to normal models, so we use `normalcdf` to find the probability of being 2.2 (or more) standard deviations below the mean. There is only about a 1.3% chance of getting 31 (or fewer) females having a BMI of more than 25 in a sample of 200, if the national rate holds here.

What should the college do about the self-reporting policy? Unless they've noticed that their female students are particularly thin, they most likely should scrap this system. Come on – how many women (or men for that matter) do you know that would *really* tell the truth about their height and weight?

Another Example

If we suppose that 13% of the population is left-handed, what's the chance that a 200-seat auditorium with 15 "lefty desks" will be able to accommodate all the lefties in a large class? Suppose there are 90 students in the class. We'd think there should be plenty of seats of each type, wouldn't we?

First, consider the conditions. We can reasonable believe that 90 students in the class are less than 10% of the population of all students at the school and could be considered a random sample, "handedness" is independent among individuals (of course, assuming no twins, triplets, etc), and we expect more than 10 lefties as well as more than 10 righties since 0.13*90 = 11.7 and 0.87*90 = 78.3. We can proceed to answer the question of interest: what's the chance of more than 15 lefties in a class of 90 students?

Since $\mu = p = 0.13$, we also have $\sigma(\hat{p}) = \sqrt{\frac{.13(.87)}{90}} = 0.0354$. Now we can do the `normalcdf` calculation to find the chance of more than 15 left-handed students, which means we'd see a sample proportion in excess of $\hat{p} = 15/90 = 0.167$. For our particular sampling distribution, 0.167 has a *z*-score of $z = \frac{.167 - .13}{.0354} = 1.05$. We see that there is about a 14.7% chance of not having enough lefty desks for the class.

```
          .1666666667
(.167-.13)/.0354
           1.04519774
normalcdf(1.05,9
9)
          .1468590807
```

SAMPLING DISTRIBUTIONS FOR A SAMPLE MEAN

The Central Limit Theorem says that as sample sizes get "large," the distribution of the sample mean $\bar{x}$ becomes normally distributed. (This is also what really happens to the sample proportion $\hat{p}$ if we consider it an average of 0's (for failures) and 1's (for successes.) There are various "rules of thumb" for how large is a large enough sample, but the real answer is that it depends on the shape of the parent (population) distribution. If the distribution is unimodal and fairly symmetric, even very small samples will give means that look normally distributed.

Did the women at the college "shave" their weights in the example above? The CDC reports that 18-year-old women in the U.S. have a mean weight of 143.74 lb with standard deviation of 51.54 lb. The 200 women in the sample reported an average weight of 140 lb. Is this unusually low, or might this just be random sampling variation at work?

We considered some of the conditions above (we can believe these women are less than 10% of the women at the "large college", and they were randomly selected.) Now, since 200 is a pretty large sample, and weights are most likely unimodal and fairly symmetric, we can also believe that the sample mean $\bar{x}$ will have a normal distribution. If these women follow the national pattern, they should have μ = 143.74 lb. The standard deviation of their average reported weights is not σ = 51.41 lb, however, but $\sigma(\bar{x}) = 51.41/\sqrt{200} = 3.64$. We find the *z*-score on this distribution that corresponds to an observed average of 140 is (140-143.74)/3.64 = -1.027.

Doing the normal calculation gives a 15.2% chance of observing an average reported weight of 140 lb or less in this sample. That's not overwhelming evidence of weight shaving. Their results have a good chance of occurring by randomness.

```
(140-143.74)/3.6
4
         -1.027472527
normalcdf(-99,-1
.027)
           .152210248
■
```

Another Example

The Centers for Disease Control and Prevention reports that the mean weight of adult men in the U.S. is 190 lb with standard deviation 59 lb. If an elevator has a limit of 10 people or 2500 lb, what's the chance that 10 men will overload it?

We can reasonably assume the 10 men are a random sample from the population, and that they represent less than 10% of the population of men who might use this elevator. It's also reasonable to believe their weights are independent of one another. Also, since weights are unimodal and fairly symmetric, it's reasonable that the average for these 10 men will have a normal distribution by the Central Limit Theorem.

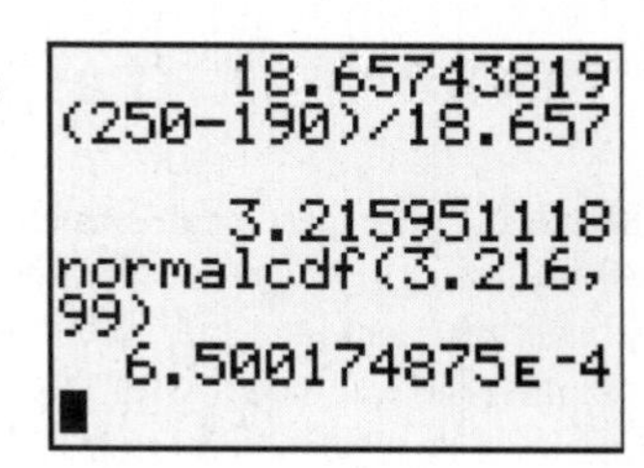

We must first realize that for the elevator to be overloaded (have total weight more than 2500 lb), the 10 men must average more than 250 lb each. If these men follow the national model, they will have $\mu = \mu(\bar{x}) = 190$ lb with $\sigma(\bar{x}) = 59/\sqrt{10} = 18.657$ lb. The *z*-score on this distribution that corresponds to an observed average of 250 is (250-190)/18.657 = 3.216. The probability of these 10 men averaging more than 250 lb is 0.00065. We have a very small chance of overloading the elevator if people follow the 10 person maximum.

WHAT CAN GO WRONG?

My Results Don't Match! (Rounding Errors)

Most "errors" are the result of rounding in intermediate steps of the problem. Rounding the standard deviations, the z-scores, etc can have an impact on the final answers. This author's advice is to carry more decimal places than is really necessary and do all rounding at the end of a series of calculations. Your instructor may have different rules. If in doubt, ASK!

My Results Don't Match! (Population vs. Sample)

The other typical error made by students in working with sampling distributions is failure to recognize the difference between dealing with one observation (or realization of a process) and a sample mean. This means that the error is usually related to having forgotten to divide the population standard deviation by $\sqrt{n}$.

Chapter 10 – Inference for Proportions

We know the sample proportion, $\hat{p}$, is normally distributed if *both* $np \geq 10$ and $n(1-p) \geq 10$. This was seen in Chapter 18 of the text and the last chapter of this manual. With this fact we found probabilities based on the normal model of obtaining certain sample proportions (or sample means). Inference asks a different question. Based on a sample, what can we say about the true population proportion? Confidence intervals give ranges of believable values along with a statement giving our level of certainty that the interval contains the true value. Hypothesis tests are used to decide if a claimed value is or is not reasonable based on the sample. (We actually did this in the last chapter – now we'll formalize it).

CONFIDENCE INTERVALS FOR A SINGLE PROPORTION

Sea fans in the Caribbean Sea have been under attack by a disease called *aspergillosis*. Sea fans that can take up to 40 years to grow can be killed quickly by this disease. In June 2000, members of a team from Dr. Drew Harvell's lab sampled sea fans at Las Redes Reef in Akumai, Mexico at a depth of 40 feet. They found that 54 of the 104 fans sampled were infected with the disease. What might this say about the prevalence of the disease in general? The observed proportion, $\hat{p} = 54/104 = 51.9\%$ is a point estimate of the true proportion, p. Other samples will surely give different results.

We can use the calculator to obtain a confidence interval for the true proportion of infected sea fans. Press STAT then arrow to TESTS. The first portion of the menu shows several hypothesis tests. We'll talk about them later. Arrowing down, we come to several possible intervals. The one we want here is choice A:1-PropZInt. Either arrow to it and press ENTER or press ALPHA MATH (A).

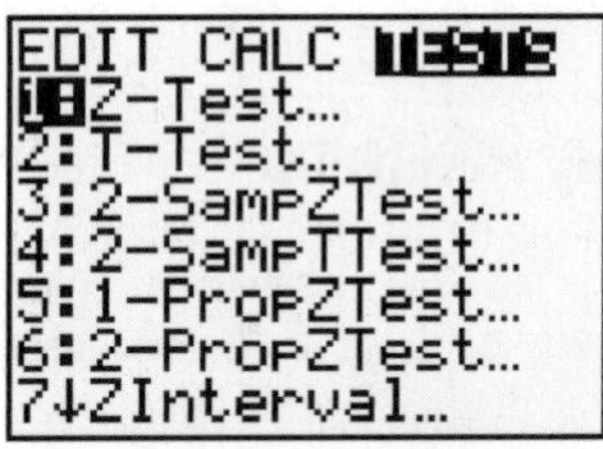

On a TI-89 all confidence intervals are located on the F7 = 2nd F2 Ints menu. The 1-PropZInt is menu option 5. Its input boxes are labeled similarly to the prompts on the 83/84 series.

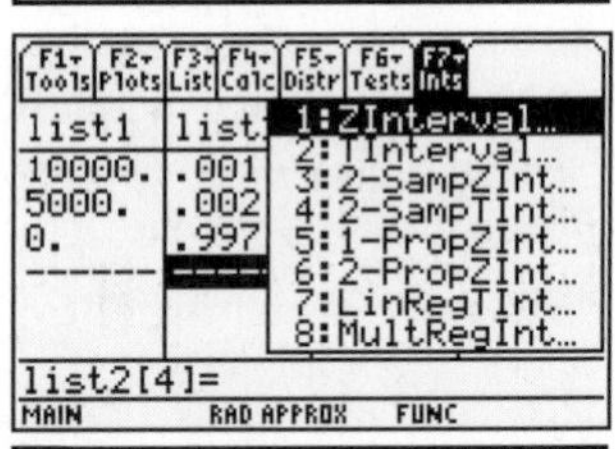

Here is the TI-83/84 input screen. Simply enter the number of observed "successes," which here is the number of infected sea fans, 54, press the down arrow, then enter the number of trials, the 104 fans that were observed, press the down arrow again to enter the desired level of confidence, finally press the down arrow again and press ENTER to calculate the results.

1-PropZInt
x:54
n:104
C-Level:.95
Calculate

How much confidence? That is up to the individual researcher. The trade-off is that more confidence requires a wider interval (more possible values for the parameter). 95% is a typical value, but the level is generally specified in each problem.

Note: TI calculators use exact critical values when finding confidence intervals rather than using the approximate 68-95-99.7 Rule.

Here are the results. The interval is 0.423 to 0.615. Remember the calculator usually gives more decimal places than are really reasonable. Usually reporting proportions to tenths of a percent is more than enough. Your instructor may give other rules for where to round final answers. The output also gives the sample proportion, $\hat{p}$, and the sample size, n.

What can we say about the prevalence of disease in sea fans? Based on this sample, we are 95% confident that between 42.3% and 61.5% of Las Redes sea fans are infected by the disease, the sample proportion is 51.9%, and the margin of error is 9.6%.

The TI-89 explicitly in its results gives the margin of error. If you are using a TI-83/84 and want to find the margin of error it is half the width of the interval. Compute $(high - low)/2$ to find it.

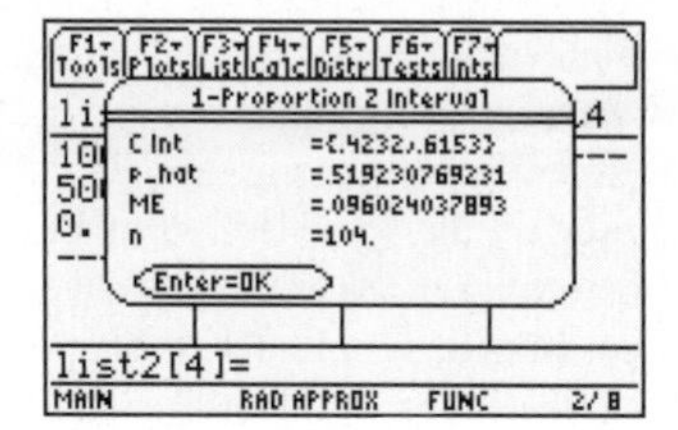

Another Example

In May 2006, the Gallup Poll asked 510 randomly selected adults the question, "Generally speaking, do you believe the death penalty is applied fairly or unfairly in the country today?" Of these, 60% answered, "Fairly." We'll build a confidence interval for the proportion of all U. S. adults who believe the death penalty is applied fairly. We had a random sample of U. S. adults which was less than 10% of the population, 60% of 510 (306) is more than 10, and 40% of 510 (204) is also more than 10, so the conditions for inference are met.

We want a 95% confidence interval. At right is the input screen. **Note:** If you are using a TI-83/84 you can actually type in the multiplication to find the number of "successes" (.6*510) and then round the result on this screen. If you are using a TI-89, you must do the multiplication on the home screen.

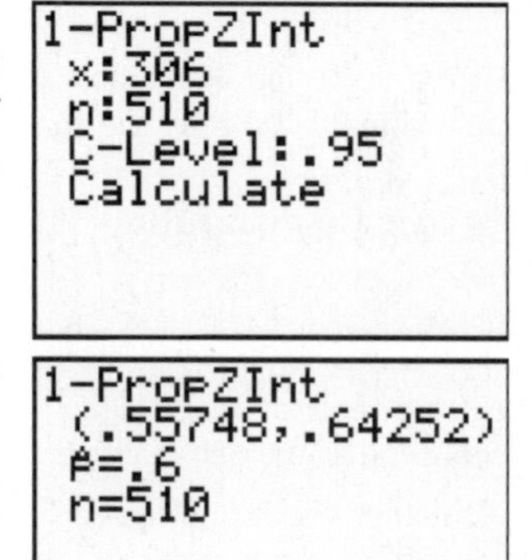

Here are the results. Based on our sample, we believe with 95% confidence that somewhere between 55.7% and 64.3% of all U. S. adults believe the death penalty is being applied fairly. Notice that since the interval only contains values above .5, we are convinced that a majority of the people in the country feels the death penalty is being applied fairly. The margin of error in these results is (.643-.557)/2 = 0.043 or 4.3%.

1-PropZInt
(.55748,.64252)
$\hat{p}$=.6
n=510

A SMALL SAMPLE CONFIDENCE INTERVAL FOR A PROPORTION

What can we do if we don't get the 10 successes and 10 failures in our sample? There is a simple adjustment to the above procedure due to Agresti/Couli (and Wilson). We merely add 2 additional "successes" and 4 "trials" to the observed data. This adjustment has been shown in numerous simulation studies to provide enough robustness for our usual procedure to actually give the desired confidence level (that is, 95% of 95% intervals *do* contain the true value.) We no longer call our estimated proportion $\hat{p}$, but $\tilde{p}$ to signify the difference.

Surgeons examined their results to compare two methods for a surgical procedure to alleviate pain on the outside of the wrist. A new method was being compared with the traditional method. Of the 45 operations with the traditional method, three were unsuccessful. What can we say about the failure rate for this type of operation? With only three failures, we can't calculate the interval as we have in the past. For these data, we find $\hat{p} = 3/45 = 6.7\%$, while using the adjustment, we find $\tilde{p} = 5/49 = 10.2\%$. Using the adjustment, we find that we are 95% confident the traditional method will fail in somewhere between 1.7% and 18.8% of cases. Be careful – the calculator has only one symbol for the "observed proportion." You must know that you used the adjusted method and properly label your results!

HYPOTHESIS TESTS FOR A SINGLE PROPORTION

Confidence intervals in general give ranges of believable values for the parameter (in this case the proportion of whatever is being termed a "success.") Hypothesis tests assess the believability of a claim about the parameter. Certainly, if a claimed value is contained in a confidence interval it is plausible. If not, it is unreasonable. Formal tests of hypotheses assess the question somewhat differently. The results given include a test statistic (here a *z*-value based on the standard Normal model) and a p-value. The p-value is the probability of an observed sample result (or something more extreme), given the claimed value of the parameter. Large p-values argue in support of the claim (your result was likely to happen by chance), small ones argue against it; in essence, if the claimed value were true the likelihood of observing what was seen in the sample is very small.

A factory casts large ingots, which are then made into structural parts for cars and planes. If they crack while being formed, the crack may get into a critical part of the final product, which compromises its integrity. Airplane manufacturers insist that metal for their planes be defect-free, so the ingot must be made over at great expense if any cracking is detected. In one plant, only about 80% of the ingots have been free of cracks. In an attempt to reduce the cracking proportion, they institute changes to the process. Since then, 400 ingots have been cast, and only 17% of them were cracked. Has the cracking rate decreased, or was the 17% just due to luck?

If we assume the 400 ingots cast under the modified process represent a random sample from all possible ingots made with this process, the conditions are for inference are reasonable (.17*400 = 68 and .83*400 = 332).

TI-83/84 Procedure

Press STAT and arrow to `TESTS`. Select choice `5:1-PropZTest` by either pressing 5 or arrowing to the selection and pressing ENTER. We are asked for p_0, the claimed value which is 20% in this case. Enter the proportion as the decimal 0.2. Then we need the number of "successes" (multiplying 0.17*400 = 68). On TI-83/84 calculators, the multiplication can be entered directly by the `x:` prompt. Press ENTER to do the computation, then press ▲ to round the result if needed. If you are using a TI-89, you must do the multiplication on the home screen. The number of trials, *n*, is 400. Then we need a direction for the alternate hypothesis (what we hope to show). This is usually obtained from the form of the question. In this case, we want to know if the proportion of cracked ingots has *decreased* which indicates we want the alternative $< p_0$. Move the blinking cursor to highlight this selection (if needed) and press ENTER to move the highlight. Finally, there are two choices for output. Selecting `Calculate` merely gives the results. Selecting `Draw` draws the normal curve and shades in the area corresponding to the p-value of the test. The TI-83/84 input screen for the test is at right.

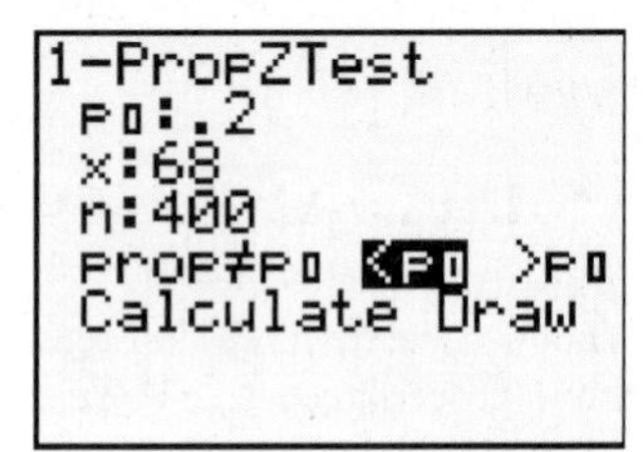

TI-89 Procedure

Hypothesis tests are located on the F6 = 2nd F1 `Tests` menu in the Statistics/List Editor Application. The input screen resembles that for the TI-83/84 series. Use the right arrow key to expand the alternate hypothesis choices and use the up or down arrows to select the one you want. Press ENTER to make the selection. You can similarly select `Calculate` or `Draw` options for the output.

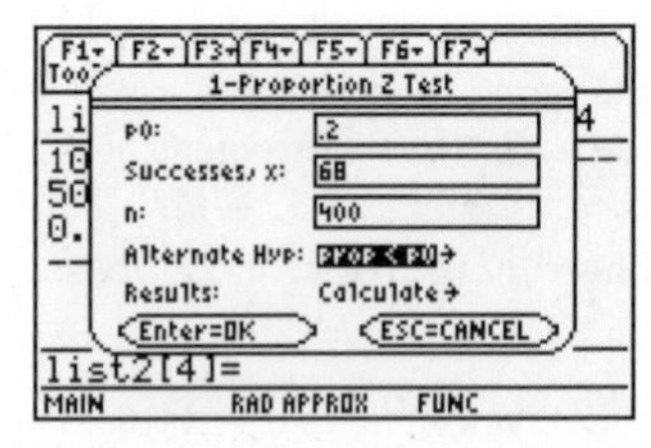

At right is the output from selecting `Calculate.` The first line of the output gives what the calculator understood the alternate hypothesis to be. Always check that this is what you intended, as it can make a difference in the p-value. The value of the test statistic is $z = -1.5$. This means that if there were no change, the observed 17% is 1.5 standard deviations above the mean. The p-value for the test is 0.067. This means that if the proportion of cracked ingots has not changed, we

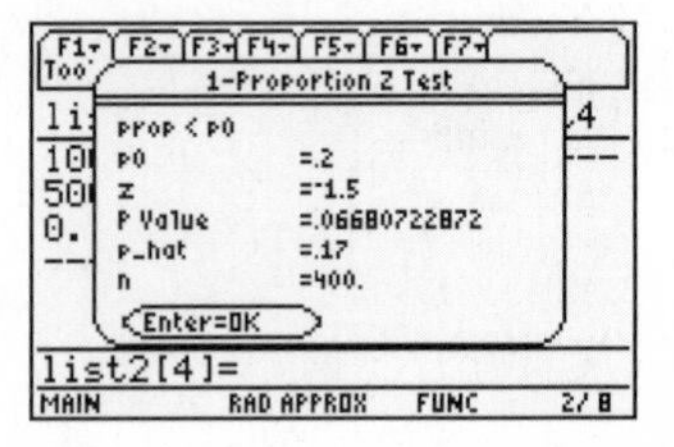

would observe a value of 17% or less only about 7 times in 100 samples of this size. Since this is not extremely rare, the factory management will have to make up their own minds about a demonstrated improvement in the process.

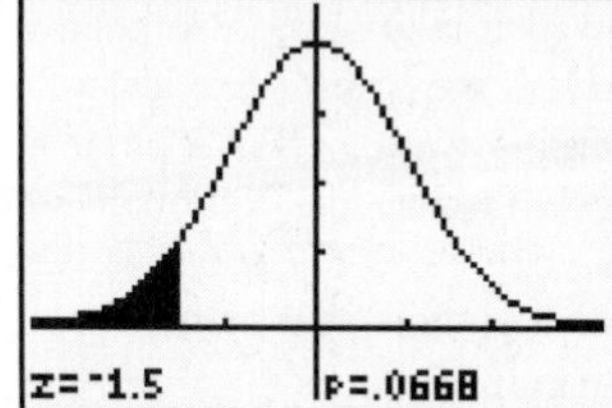

Here is the output when DRAW is selected. The test statistic and p-value are given. The area under the standard normal curve corresponding to the p-value is also shaded. Since the alternate hypothesis was $p < .20$, only area below the calculated test statistic corresponds to the p-value.

Another Example

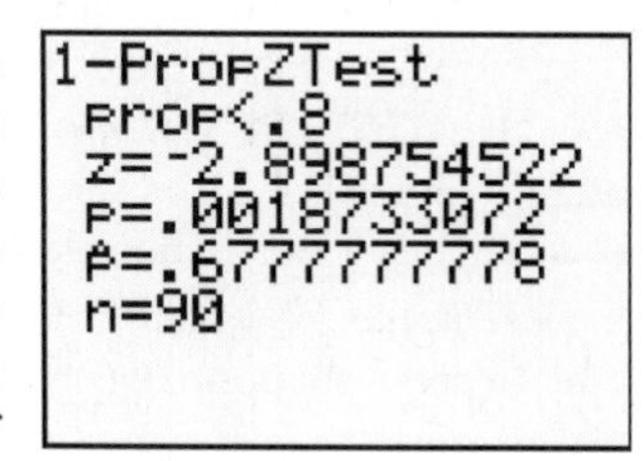

The Department of Motor Vehicles claimed that over 80% of candidates pass the driving test, but a newspaper reporter's survey of 90 randomly selected local teens who had taken the test found only 61 had passed. The observed sample statistic is $\hat{p} = 61/90 = 0.678$. This is lower than the claimed passing rate. Is it enough lower to cast doubt on the claim made by the DMV? The teens were randomly selected, it's reasonable that they're independent of one another, they do represent less than 10% of teens who might take the test, and we have more than 10 each of failures and successes.

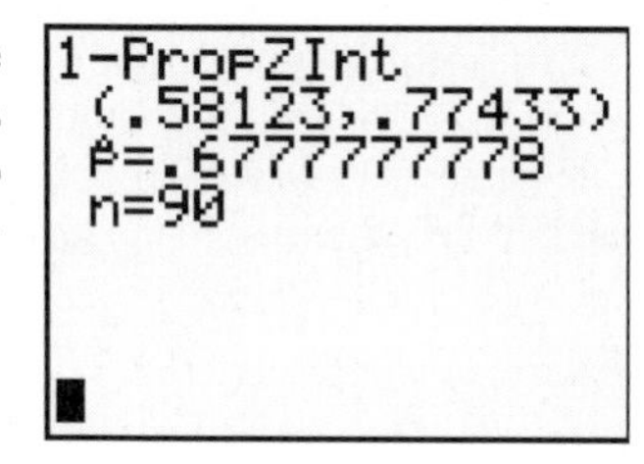

Here are the results. The z-statistic of -2.899 means that if the DMV is correct, the observed 67.8% is almost 3 standard deviations below the mean, which is extremely rare. The p-value is 0.0019. If the DMV were correct, we'd expect to see our sample results only about twice in 1000 samples. This extremely small p-value argues that these data show the DMV is wrong (at least for teen drivers.) What is the passing rate for teen drivers? Based on these data, we are 95% confident between 58.1% and 77.4% of teen drivers will pass the test. Notice that since the high end of the confidence interval is below 80%, this also supports our conclusion that the DMV's claim is wrong.

Yet another example

In some cultures, male children are valued more highly than females. In some countries with the advent of prenatal tests such as ultrasound, there is a fear that some parents will not carry pregnancies of girls to term. A study in Punjab, India[1] reports that in 1993 in one hospital 56.9% of the 550 live births were males. The authors report a baseline for this region of 51.7% male live births. Is the sample proportion of 56.9% evidence of a change in the percentage of male births?

The baseline proportion of males for the area is 51.7% in this case. This has been entered in decimal form as 0.517. Then we need the number of "successes" (Multiplying 0.569*550 gives 312.95 which rounds to 313). The multiplication can be entered directly by the x: prompt, then press [ENTER], then up arrow to round the result if needed. The number of trials, n, is 550. Then we need a direction for the alternate hypothesis (what we hope to show). In this case, we want to know if the proportion has changed which means we select the $\neq p_0$ alternative.

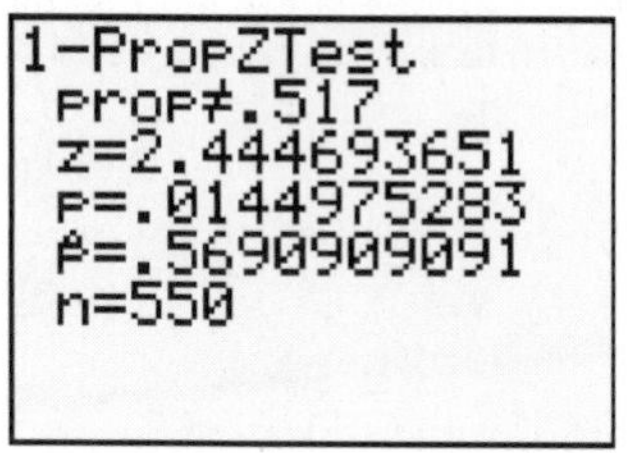

The value of the test statistic is z = 2.44. This means that if there were no change, the observed 56.9% is 2.44 standard deviations above the mean. The p-value for the test is 0.0145. This means that if the proportion of male births is still 51.7%, we would observe a value of 56.9% or greater only about 1 times in 100 samples. Since this is very rare, we will reject the null hypotheses and conclude that we believe that, based on these data, the true proportion of male births in Punjab is now greater than the baseline 51.7%.

[1] "Fetal Sex determination in infants in Punjab, India: correlations and implications," E.E. Booth, M. Verna, R. S. Beri, *BMJ*, 1994; 309:1259-1261 (12 November).

CONFIDENCE INTERVALS FOR THE DIFFERENCE OF TWO PROPORTIONS

Do men take more risks than women? A recent seat belt study in Massachusetts found that, not surprisingly, male drivers wear seat belts less often than women, but that men's belt-wearing jumped more than 16 percentage points when a woman was a passenger[2] Seat belt use was recorded at 161 locations using random sampling methods developed by the National Highway Traffic Safety Administration (NHTSA). Of 4,208 male drivers with female passengers, 2,777 (66.0%) wore seat belts. But among the 2,763 male drivers with male passengers only 1,363 (49.3%) wore seat belts. What do we estimate the true gap in seat belt use to be? We want a confidence interval for the difference in the true proportions, $p_W - p_M$. If this interval contains 0, there is no statistical evidence of a difference. Note that the order makes no real difference, but since observed seat belt usage is higher with the woman passenger, most people tend to prefer positive differences.

From the STAT, TESTS menu select choice B:2-PropZInt. This is menu option 6 on the [F7] Ints menu on a TI-89. The calculator uses groups 1 and 2, not men and women, and calculates results based on $group1 - group2$. Decide (and keep note of) which number you assign to each group. Since it is usually easier to deal with positive numbers, we will call the women passengers group 1.

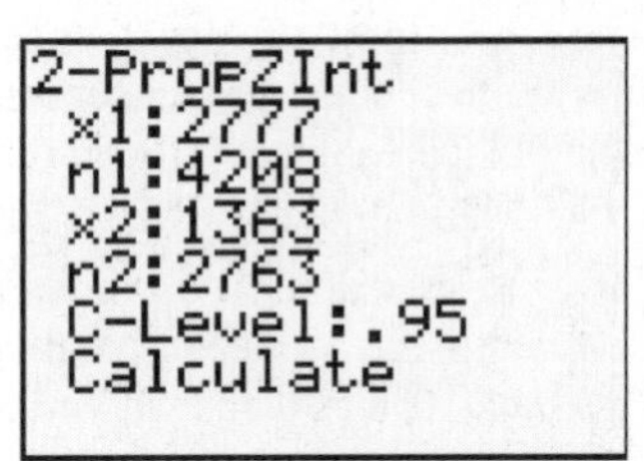

The calculated interval is 0.143 to 0.190. This means we are 95% confident, based on this poll, the proportion of men who buckle up with a woman passenger is between 14.3% and 19.0% more than the proportion of men who buckle up with a male passenger. Since 0 is not in the interval, there is a definite difference in male seat belt behavior due to the gender of the passenger (at least in Massachusetts.)

```
2-PropZInt
 (.14313,.19013)
 p̂1=.6599334601
 p̂2=.4933043793
 n1=4208
 n2=2763
```

Note: A word of caution about these intervals: when looking at the results one must be clear which group was labeled Group 1 and which was Group 2. If I had called the male passengers Group 1 in the example above, the numerical results would be the same, but each end of the interval would be negative. In that case the interpretation would be that male drivers were between 14.3% and 19.0% *less* likely to buckle up with a male passenger than with a female.

HYPOTHESIS TESTS FOR A DIFFERENCE IN PROPORTIONS

The National Sleep Foundation asked a random sample of 1010 U.S. adults questions about their sleep habits. The sample was selected in the fall of 2001 from random telephone numbers.[3] Of interest to us is the difference in the proportion of snorers by age group. The poll found that 26% of the 184 people age 30 or less reported snoring at least a few nights a week; 39% of the 811 people in the older group reported snoring. Is the observed difference of 13% real or merely due to sampling variation?

The null hypothesis is that there is no difference, which means $p_1 = p_2$ or $p_1 - p_2 = 0$. (The calculator and most computer statistics packages can only test assumed differences of 0; if there were an assumed difference, such as the belief that older people had 10% more snorers than young people, one would need to compute the test statistic "by hand.")

---

[2] Massachusetts Traffic Safety Research Program (June 2007)

[3] 2002 *Sleep in America Poll*, National Sleep Foundation. Washington D.C.

Decide which group will be Group 1. We will use the older people as Group 1. (There will be no difference in the results, but again, positive numbers are generally easier for most people to deal with.) From the STAT TESTS menu select choice 6:2-PropZTest. The number of snorers in the older group was 0.39*811 = 316.29 (rounded to 316); for the younger group the number of snorers was 0.26*184 = 47.84 which rounds to 48. The chosen alternate is $\neq p2$ since we just want to know if there is a difference.

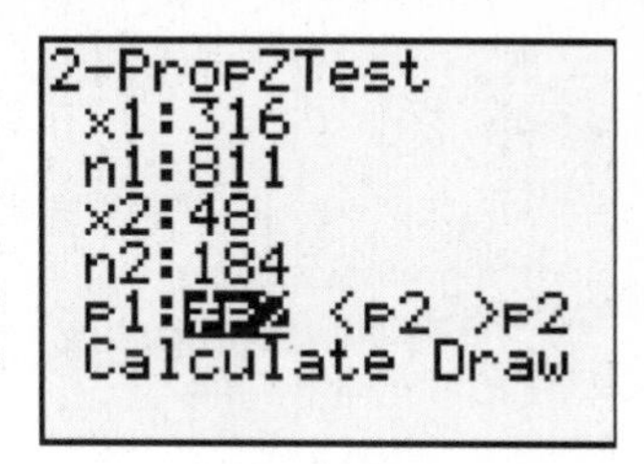

Here are the results. Be careful here: there are lots of p's floating around. We see the chosen alternative, $p1 \neq p2$, The test statistic is z =3.27, which means that if there were no difference in the proportion of snorers the observed difference (about 13%) is more than 3 standard deviations above the mean. The p-value for the test is given as $p = 0.0011$. The next values given are the observed proportions in each group, $\hat{p}_1$ and $\hat{p}_2$ then an overall $\hat{p}$ which represents the observed proportion, *if there were no difference in the groups*.

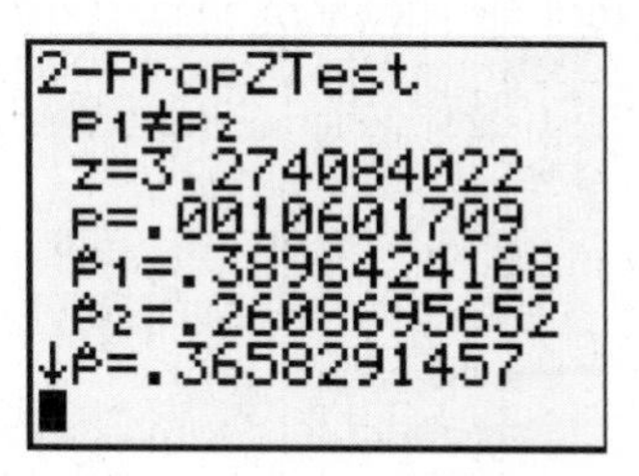

One can arrow down to see the sample sizes for the two groups as well. Since the p-value for the test is so small, we believe there is a difference in the rate of snorers based on this poll. We can further say the proportion of snorers is greater in older people than in those under 30 – their observed proportion (39.0%) is larger than for the observed proportion (26.1%) for the younger group.

How big is the difference? We are 95% confident, based on this data the proportion of snorers in older adults is between 5.7% and 20.1% larger than for those under 30.

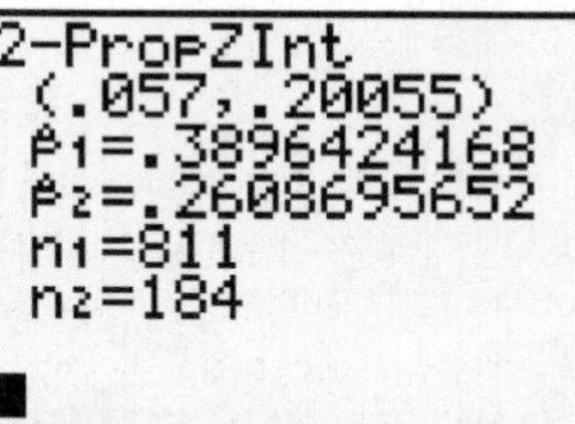

WHAT CAN GO WRONG?

Err: Domain?

This error stems from one of two types of problems. Either a proportion was entered in a 1-PropZTest which was not in decimal form or the numbers of trials and/or successes was not an integer. Go back to the input screen and correct the problem.

Err:Invalid Dim?

This can be caused by selecting the DRAW option if another Statistics plot is turned on. Either go to the STAT PLOT menu (2nd Y=) and turn off the plot or redo the test selecting CALCULATE.

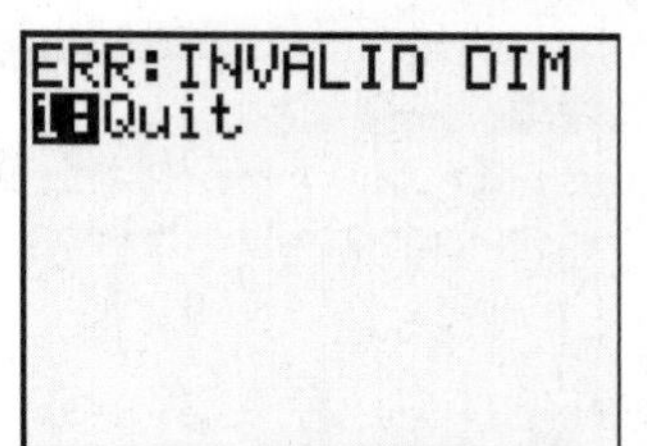

What are the weird lines?

This is caused (as usual in graphing errors) by having an equation entered on the Y= screen or a STAT PLOT turned ON. Remember, the calculator always tries to plot everything it knows about. Clear the Y= screen, and turn STAT PLOTS off.

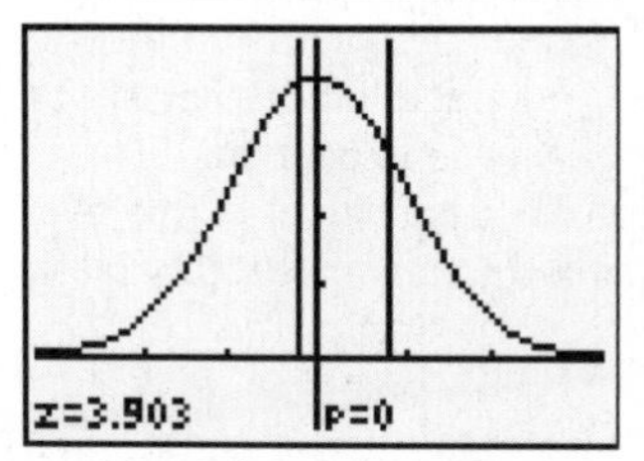

Bad Conclusions.

Small p-values for the test argue against the null hypothesis. If the p-value is small, one rejects the null hypothesis and believes the alternate is true. If the p-value is large, the null hypothesis is not rejected; this does *not* mean it is true – we simply haven't gotten enough evidence to show it's wrong. Be careful when writing conclusions to make them agree with the decision.

Chapter 11 – Inference for Means

Inference for means is a little different than that for proportions. Most introductory statistics texts base this on standard normal models, which is truly appropriate only if the population standard deviation, σ, is known. In most cases this is not true; the only time one might really believe σ is known is in the case of quality control sampling where a production line has been tracked for a long time. If σ is not known confidence intervals and hypothesis tests should be based on *t* distributions. T distributions have larger critical values (multipliers in confidence intervals) than the standard normal curve to allow for the additional uncertainty in having estimated two parameters of the population – the mean and standard deviation, instead of just one (as with proportions). These distributions become the standard normal distribution when the sample size is very large (infinite).

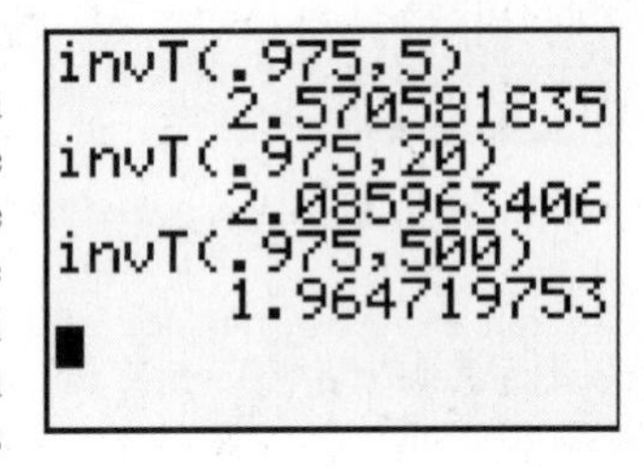

On the TI-84 or 89, we can see the impact of sample size on these critical values. From the **DISTR** menu ([2nd][VARS], or [F5] on a TI-89), choose **4:invT(**. As with **InvNorm** which is used to find percentiles of the normal distribution, the parameters for this command are the area to the left of the desired point and the degrees of freedom ($n - 1$ for a single sample). For a 95% confidence interval, we saw before that z^* is 1.96. The *t* critical values at right are also for a 95% interval (if there is 95% in the middle of the curve, there is 97.5% to the left of the high end of the region) and represent samples of size 6, 21, and 501. Notice that as degrees of freedom or sample size get larger, these numbers get closer to 1.96.

Small sample sizes give rise to their own problems. If the sample size is less than about 30, the Central Limit Theorem does not apply, and one cannot merely assume the sample mean has a normal distribution. In the case of small samples, you must check that the data come from a (at least approximately) normal population, usually by normal probability plots or boxplots since histograms are not useful with small samples.

CONFIDENCE INTERVALS FOR A MEAN

Residents of Triphammer Road are concerned over vehicles speeding through their area. The posted speed limit is 30 miles per hour. A concerned citizen spends 15 minutes recording the speeds registered by a radar speed detector that was installed by the police. He obtained the following data:

29	34	34	28	30	29	38	31	29	34	32	31
27	37	29	26	24	34	36	31	34	36	21	

We want to estimate the average speed for all cars in this area, based on the sample. Enter the data in a list. Here, I have entered them into list **L1**. This is a small sample – there are only 23 observations, so we should check to see if the data looks approximately normal.

L1	L2	L3 2
29		------
34		
34		
28		
30		
29		
38		

L2(1)=

A normal plot of the data looks relatively straight, with no outliers, so it's reasonable to continue. This plot shows some granularity (repeated measurements of the same value) but no overt skewness or outliers. If you've forgotten how to create normal probability plots, return to Chapter 4 of this manual.

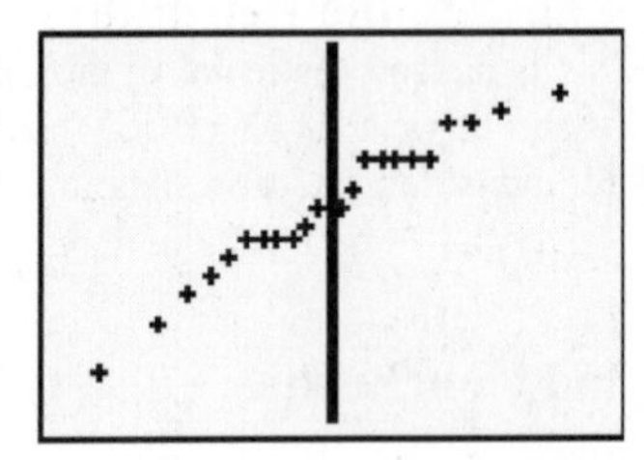

TI-83/84 Procedure

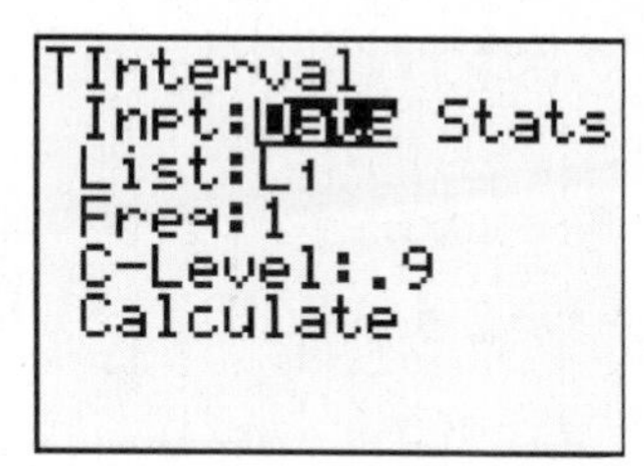

Press [STAT], arrow to TESTS then select choice 8:TInterval. You have two choices for data input: using data in a list such as we have or inputting summary statistics from the sample. Move the cursor to DATA and press [ENTER] to move the highlight. Enter the name of the list with the data ([2nd][1] for L1). Each observation occurred once, so leave Freq as 1. If there were a separate list of frequencies for each data value, that would be entered here. Enter the desired amount of confidence (here, 90%, but in decimal form) and finally press [ENTER] to perform the calculation.

TI-89 Procedure

From the Statistics/List editor, press [2nd][F2] =[F7] (Ints). Select choice 2:TInterval. You have two choices for data input: using data in a list such as we have or inputting summary statistics from the sample. Pressing the right arrow allows you to make the selection. Press [ENTER] to get the next input screen. Enter the name of the list with the data ([2nd][-] takes you to the [VAR-LINK] screen). Each observation occurred once, so Freq should be 1. If there were a separate list of frequencies for each data value, that would be entered here. Enter the desired amount of confidence (here, 90%, but in decimal form) and finally press [ENTER] to perform the calculation.

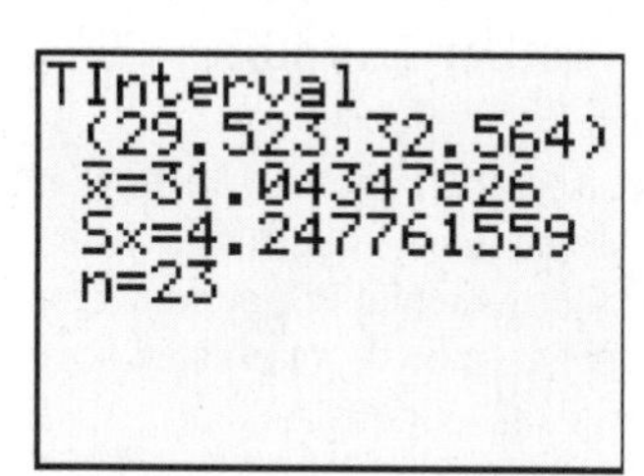

Here are the results. Based on this sample, we are 90% confident the average speed for all cars on this road is between 29.5 and 32.6 miles per hour. There are two caveats here: the first is that this was not a truly random sample but a convenience one (only one 15 minute period was sampled). Also, the presence of the radar speed detector may have influenced the drivers at that time. Drivers may be driving over the posted 30 miles per hour limit, but since 30 is included in the interval, we have not shown conclusively that drivers are speeding, on average.

What about the extra decimal places? The general rule here, as with reporting means and standard deviations in general, is to report one more decimal place than in the original data. Our data were in integer miles per hour, so report one decimal place. Your instructor may have a different rule for this, so please listen to him or her.

What if we don't have the data? In the case of a small sample size, one must assume the data comes from an approximately normal population. If the sample is "large," the Central Limit Theorem will apply and $\bar{x}$ will be approximately normal.

Another Example

In 2004 a team of researchers published a study of contaminants in farmed salmon. Fish from many sources were analyzed for several contaminants, one of which was the insecticide mirex. After outliers from one particular farm were removed, the remaining 150 fish averaged 0.0913 ppm with stancard deviation s = 0.0495 ppm. What is a 95% confidence interval for the mean mirex contamination? Here we have moved the highlight from Data to Stats. When this is done, the input screen changes to ask for the sample mean, standard deviation, sample size and confidence level.

Pressing [ENTER] to calculate the interval tells us we are 95% confident, based on this sample the mean mirex contamination in farm-raised salmon is between 0.083 and 0.099 ppm.

A ONE SAMPLE TEST FOR A MEAN

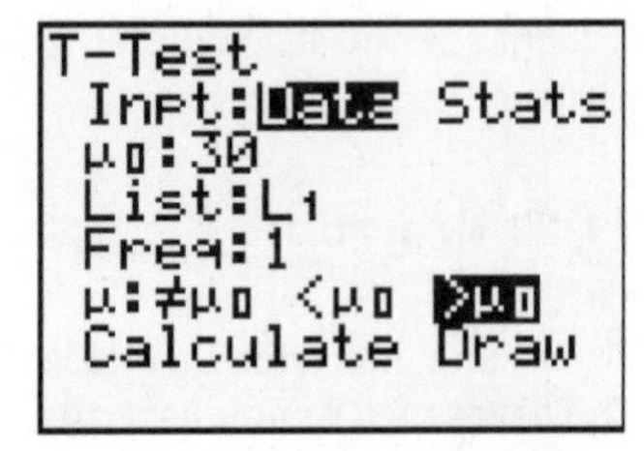

We can also do a hypothesis test to decide whether the mean speed is more than 30mph. From the `STAT TESTS` menu, select choice `2:Ttest`. We are still using data in list `L1`. μ_0 is set to 30 since that's the speed limit we're comparing against. The alternate has been selected as $> \mu_0$ since we want to know if people are going too fast, on average. Notice we have the options of `Calculate` and `Draw` here, just as we did on tests of proportions.

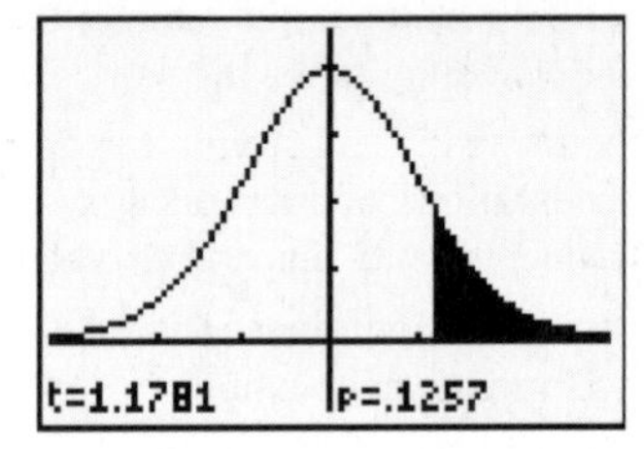

Selecting `Draw` yields the screen at right. We can clearly see the shaded portion of the curve which corresponds to the p-value for the test of 0.1257. The calculated test statistic is t = 1.178. The p-value indicates we'll expect to see a sample mean of 31.04 (the mean from our sample) or higher by chance about 12.5% of the time by randomness when the mean really is 30. That's not very rare. We fail to reject the null and conclude these data do not show motorists on the street are speeding on average.

Another Example

Researchers tested 150 farm-raised salmon for organic contaminants. They found the mean concentration of the carcinogenic insecticide mirex to be 0.0913 parts per million (ppm), with standard deviation 0.0495 ppm. The Environmental Protection Agency's recommended "screening value" for mirex is 0.08 ppm. Do farm-raised salmon appear to be contaminated beyond the level permitted by the EPA?

The salmon were randomly selected, and raised and purchased in many places, so they should be independent of each other.. Further, these represent a really small fraction of the potential salmon available for sale. With a sample size of 150, the actual shape of the distribution is of small concern (it's actually somewhat right skewed, with no outliers.) Since all the conditions are met, we may proceed to the test. We want to know if these salmon appear to have contaminant levels that exceed the EPA permitted, so the form of the alternate hypothesis is "$>\mu_0$."

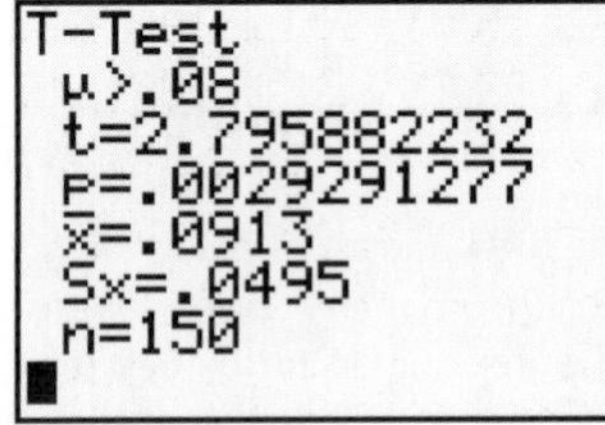

We see on the results screen that if the level were indeed .08 (or less) that the observed mean of 0.0913 is 2.80 standard deviations above that level. The probability of our observed sample mean or something higher is 0.0029. This is an extremely small p-value, so we have very strong evidence that these fish do indeed exceed the EPA screening value. One might want to think twice about eating farm-raised salmon.

COMPARING TWO MEANS – CONFIDENCE INTERVALS

Should you buy name brand or generic batteries? Generics cost less, but if they do not last as long on average as the name brand, spending the extra money for the name brand may be worthwhile. Data were collected for six sets of each type of battery, which were used continuously in a CD player until no more music was heard through the headphones. The lifetimes (in minutes) for the six sets were:

Brand Name:	194.0	205.5	199.2	172.4	184.0	169.5
Generic:	190.7	203.5	203.5	206.5	222.5	209.4

The first step in performing a comparison such as this one (or any!) should always be to plot the data. Here, a side-by-side boxplot is natural.

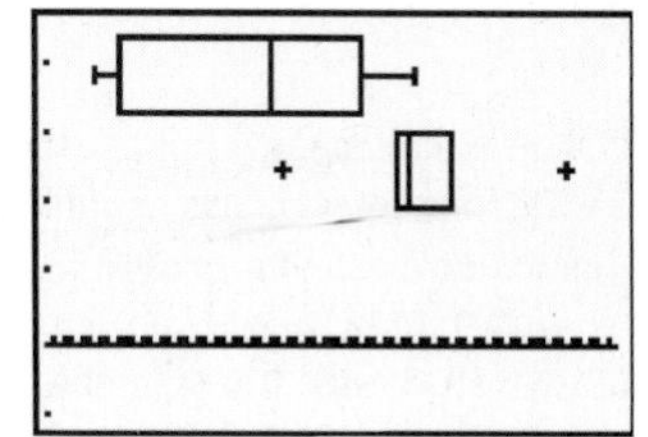

We have entered the data into list L1 for the Brand name batteries, and list L2 for the generics. We defined two boxplots to identify outliers on the STAT PLOT menu ([2nd][Y=]). For more on these plots, see Chapter 3 of this manual. From the plot, it certainly appears the generics last longer than the name brand batteries; they also seem more consistent (they have a smaller spread). There are two outliers for the generic batteries, but with a sample size this small the outlier criteria are not very reliable. Neither of the extreme values are unreasonable, so it's safe to continue.

From the STAT TESTS menu select choice 0:2-SampTInt (option 4 on the [F7] Ints menu on a TI-89). Our data are already entered, so move the highlight (if necessary) to Data and press [ENTER]. The data were in lists L1 and L2, and each value in the lists occurred once. The confidence level has been set to 95% (entered as always in decimal form). The next option is new. Pooled: refers to whether the two groups are believed to have the same standard deviation. Visually, this is not true for our two battery samples. In general, unless there is some reason to believe the groups have the same spread, it's safest to answer this question with No. Reasoning behind this question has to do with computing a "pooled standard deviation" (or not) and the number of degrees of freedom for the test. Before the advent of computers (and statistical calculators) there were many recipes for handling this question, since the calculation of degrees of freedom in the unpooled case is complex. Luckily, we just let the calculator do the work.

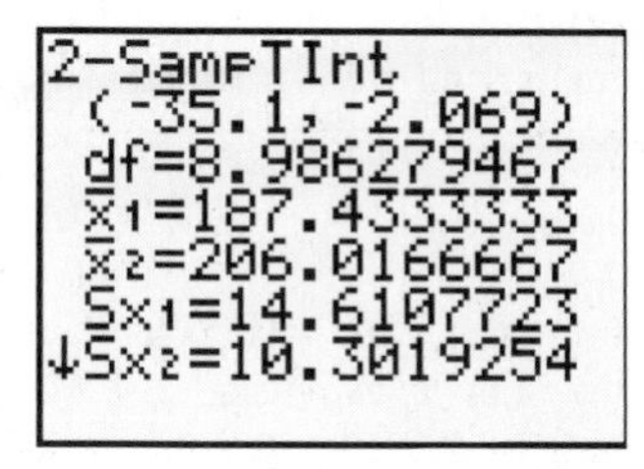

Pressing [ENTER] to calculate the interval gives the screen at right. We see we are 95% confident the average life of the name brand batteries is between 35.1 and 2.1 minutes *less* than the average life of the generic batteries. (Remember, it's always *group*1 – *group*2 in the interval, just as we found with proportions). The next line gives the degrees of freedom for the interval – notice they're not even integer-valued. We also see the two sample means and standard deviations. The ↓ at the bottom left indicates more output can be obtained (the sample sizes). Assuming generic batteries are cheaper than name brand ones, it certainly would make sense to buy them.

The Subtly Refilling Soup Bowl

Do people take visual or internal cues when they eat? Researchers wanted to examine this question. Twenty-seven people were each assigned randomly to eat soup – one group from regular bowls, and the other to eat soup from bowls that were secretly refilled. Which group ate more? How much more? The results of the experiment are summarized in the table below.

	Ordinary bowl	Refilling Bowl
n	27	27
$\bar{y}$	8.5 oz.	14.7 oz
s	6.1 oz.	8.4 oz.

It appears that the people with the refilling bowl are more, but is the difference statistically significant? What does a 95% confidence interval say about the difference? In the screen at right, I have entered the summary statistics given above.

```
2-SampTInt
 (-10.22,-2.182)
 df=47.45553479
 x̄1=8.5
 x̄2=14.7
 Sx1=6.1
↓Sx2=8.4
```

Group 1 was the group with the ordinary bowl. We notice that both ends of the confidence interval are negative. This means that we are 95% confident based on this experiment that people with ordinary bowls will eat between 2.18 and 10.22 oz. *less* than people with the refilling bowl. It would seem that fullness of the bowl (rather than the stomach) is the more important cue.

TESTING THE DIFFERENCE BETWEEN TWO MEANS

If you bought a used camera in good condition, would you pay the same amount to a friend as to a stranger? A Cornell University researcher wanted to know how friendship affects simple sales such as this.[1] One group of subjects was asked to imagine buying from a friend whom they expected to see again. Another group was asked to imagine buying from a stranger. Here are the prices offered.

Friend	$275	300	260	300	255	275	290	300
Stranger	260	250	175	130	200	225	240	

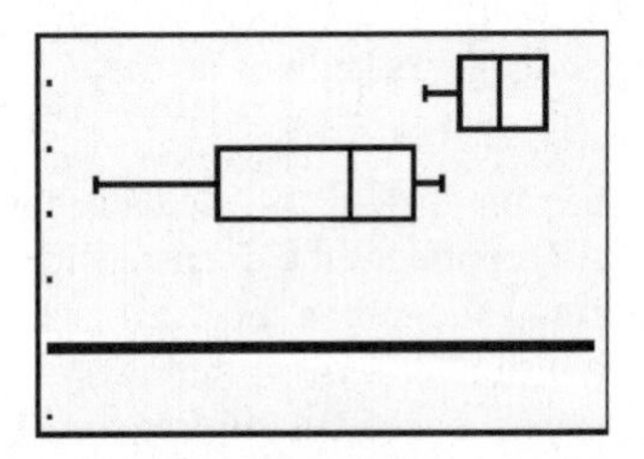

Here are side-by-side boxplots of the data. There certainly looks to be a difference. Prices to buy from strangers seem lower and much more variable than the prices for buying from a friend. As with the battery example, looking at skewness or outliers for these small samples is difficult, but the plots look reasonable.

From the `STAT TESTS` menu, select choice `4:2-SampTTest`. Again, we have the data in two lists, so `Data` is highlighted as the input mechanism, we have indicated the data are in lists `L1` and `L2`, and each data value has a frequency of 1. The alternate hypothesis is $\mu_1 \neq \mu_2$ since our original question was "would you pay the same amount." Again, we have indicated `No` in regards to pooling the standard deviations (the spreads of the distributions do not look equal and there is no reason to believe they should be the same).

[1] Halpern, J.J. (1997). The transaction index: A method for standardizing comparisons of transaction characteristics across different contexts, *Group Decision and Negotiation*, 6(6), 557-572.

Pressing ENTER to calculate the test gives the screen at right. The computed test statistic is t = 3.766, and the p-value is 0.006. From these data we conclude that not only are people not going to pay the same amount to a friend as to a stranger, they're willing to pay more. We might even go so far as to warn people not to pay *too much* to friends.

```
2-SampTTest
 μ1≠μ2
 t=3.766049006
 p=.0060025794
 df=7.622947934
 x̄1=281.875
↓x̄2=211.4285714
```

Back to the Soup

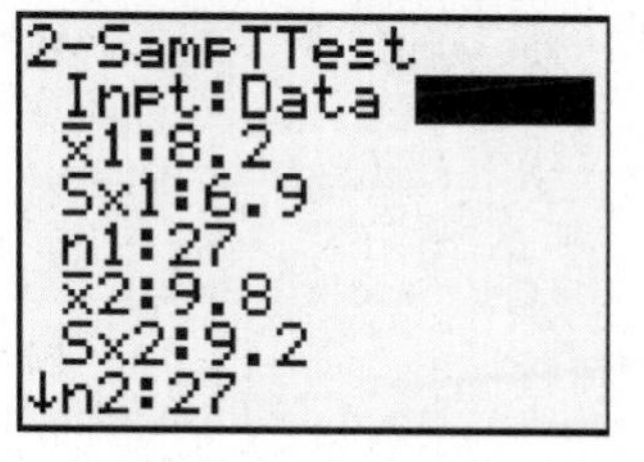

This manual (and the authors of your text) have said that unless there is some reason to believe the spreads of the two samples should be the same, it's best to use the non-pooled test. There are occasions where answering "Yes" to the pooled question makes sense. The individuals in the soup experiment not only had the actual amount eaten measured, but they were asked how much they thought they had eaten. If the two groups really were equivalent before the soup experiment, they should have similar standard deviations, and in fact, the standard deviation for the ordinary bowl was 6.9 oz and the standard deviation for the refilling bowl was 9.2 oz. We want to examine the question of whether or not there is a significant difference in the amount of soup the people thought they'd eaten. Since the two groups should have similar spreads (and the statistics are similar), this is a good argument to use a pooled standard deviation with these data.

The results indicate no significant difference in the amount the subjects thought they'd eaten. The p-value for the test is 47.3%. Notice the degrees of freedom here are different from those used in the confidence interval.

```
2-SampTTest
 μ1≠μ2
 t=-.7229429458
 p=.4729542179
 df=52
 x̄1=8.2
↓x̄2=9.8
```

PAIRED DATA

The two sample problems considered above used two *independent* samples. Many times data which might seem to be for two samples are naturally paired (say, examining the ages of married couples – each couple is a natural pair) or are even two observations on the same individuals. In such cases one works with the *differences* in each pair, and not the two sets of observations. The reason for this is to eliminate variability among the pairs and focus on the difference within the pairs.

Do flexible schedules reduce the demand for resources? The Lake County (IL) Health Department experimented with a flexible four-day week. They recorded mileage driven by 11 field workers for a year on an ordinary five-day week, then they recorded the mileage for a year on the four-day week.[2] The data are below. The first important fact to realize is that we have data on the same individuals under the two different schedules. These are *not* independent samples, but rather *paired data*.

Name	5 day mileage	4 day mileage
Jeff	2798	2914
Betty	7724	6112
Roger	7505	6177
Tom	838	1102
Aimee	4592	3281
Greg	8107	4997
Larry G	1228	1695
Tad	8718	6606
Larry M	1097	1063
Leslie	8089	6392
Lee	3807	3362

[2] Catlin, Charles S. Four-day Work Week Improves Environment, *Journal of Environmental Health*, Denver, March 1997 **59**:7.

Cursory examination reveals that after the change, some drove more, and some less. It is also easy to see there are large differences in the miles driven by the different workers. It is this variation between individuals that paired tests seek to eliminate.

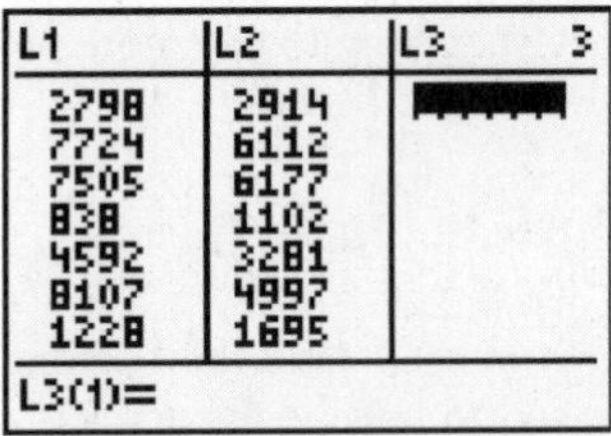

We have entered the data into the calculator; the five-day week mileages are in list L1, and the four-day mileages are in list L2.

We need to find the differences. On the home screen, one could press [2nd][1][−][2nd][2][STO▸][2nd][3] which results in the command L1 – L2 →L3. However, since we are in the editor, an easier way is to move the cursor to highlight the name of an empty list and enter the command.

The command will look as at right. On a TI-89, the command is the same, with the exception of using [VAR-LINK] to access list names. Pressing [ENTER] to complete the calculation will display the first few values. If you want to see the entire list, scroll through it using the down and up arrows.

L1	L2	L3
2798	2914	------
7724	6112	
7505	6177	
838	1102	
4592	3281	
8107	4997	
1228	1695	

L3 =L1−L2

We need to check if the *differences* are approximately normal (or certainly at least have no strong skewness or outliers). For this type of test, we are using the differences as the data so the Nearly Normal condition applies to them and not the original data. We define the normal plot as at right to use the differences which were just created. Press [ZOOM][9] to display the plot.

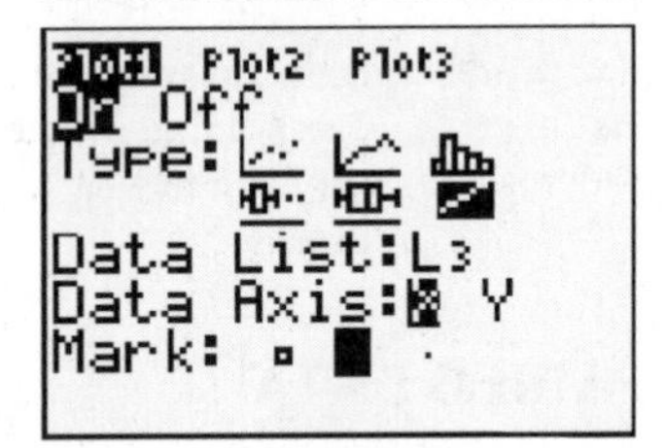

The plot at right is not perfectly straight. However, there are no large gaps, so no extreme outliers.

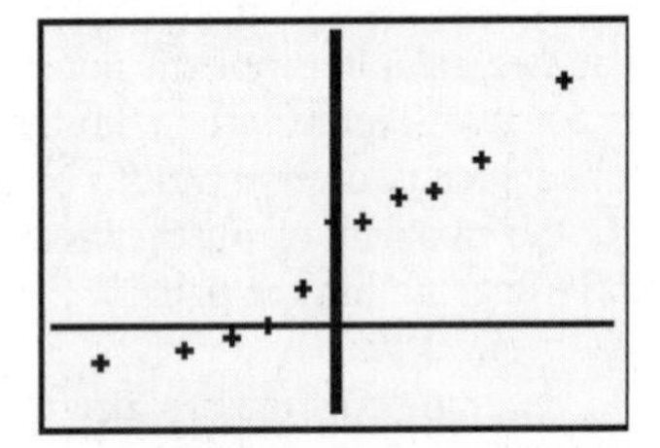

We now proceed to the test. We will perform a one-sample test using the differences as the data. From the STAT TESTS menu, select 2:T-Test. If the change in work week made no difference, the average value of the computed differences should be 0, so this is the value for μ_0. We are using the data from list L3 as the input, and have selected the alternative hypothesis as $\mu \neq \mu_0$.

Pressing [ENTER] when the cursor is over Calculate displays the results. The computed test statistic is $t = 2.85$ and the p-value is 0.017. We conclude that these data do indicate a difference in driving patterns between a five-day work week and a 4-day work week. Further since the average difference is positive (982 miles) it seems that employees drove less on the four-day week (the subtraction was five-day – four-day mileages). It's hard to say if the difference is meaningful to the department (remember, statistical significance is not necessarily practical significance). If so, they may want to consider changing all employees to four-day weeks.

```
T-Test
 μ≠0
 t=2.858034426
 p=.0170141282
 x̄=982
 Sx=1139.568339
 n=11
```

We can go further and compute a confidence interval for the average difference. Select 8:TInterval from the STAT TESTS menu, and define the interval as at right.

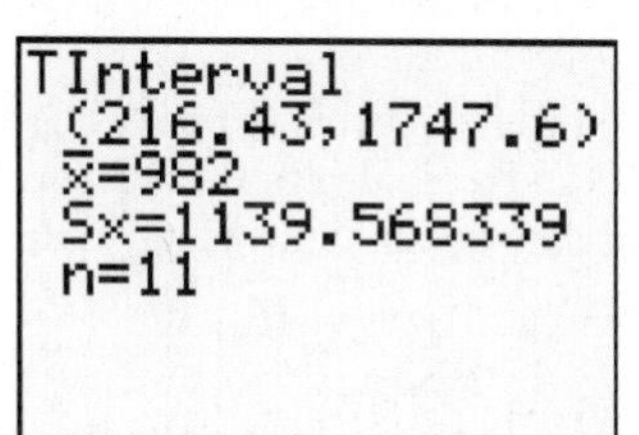

Pressing ENTER to calculate the interval, we find we are 95% confident the five-day work week will average between 216.4 and 1747.6 more yearly miles than a 4-day work week.

Speed Skaters

In the 2006 Olympics, there were allegations that the outer lane in the speed skating competition was "faster" than the inner lane. The racers are randomly assigned to race in pairs throughout the day of competition, and conditions may vary as the day goes on. Even though the skaters switch lanes halfway through the race, it was believed that those who started in the outer land had an advantage. Is this so? The table below gives the times (and differences) for each pair of skaters in the women's 1500m race, according to which lane was the first.

Skating Pair	Inner Time	Outer Time	Difference
1	125.75	122.34	3.41
2	121.63	122.12	-0.49
3	122.24	123.35	-1.11
4	120.85	120.45	0.40
5	122.19	123.07	-0.88
6	122.15	122.75	-0.60
7	122.16	121.22	0.94
8	121.85	119.96	1.89
9	121.17	121.03	0.14
10	124.77	118.87	5.90
11	118.76	121.85	-3.09
12	119.74	120.13	-0.39
13	121.60	120.15	1.45
14	119.33	116.74	2.59
15	119.30	119.15	0.15
16	117.31	115.27	2.04
17	116.90	120.77	-3.87

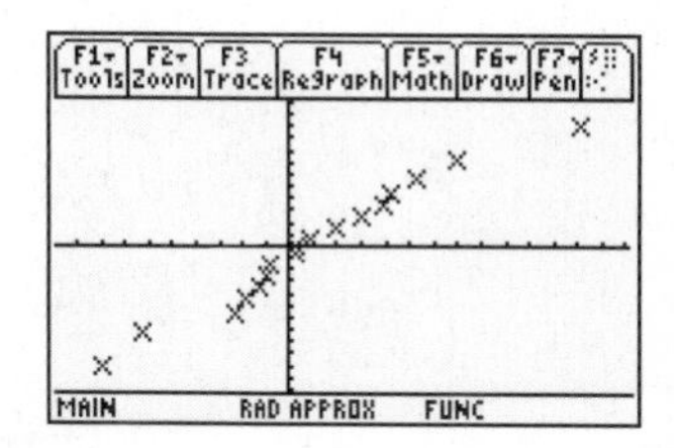

The normal plot at right of the differences shows no overt skew or outliers. A histogram of the differences also appears relatively symmetric.

Here is my input screen for the test using the TI-89. The differences are in `list1`, and our question of interest is whether (or not) the two lanes are different, so the alternate is "not equal." Again, if the two lanes are equivalent (fair), the average difference in the pairs times should be 0.

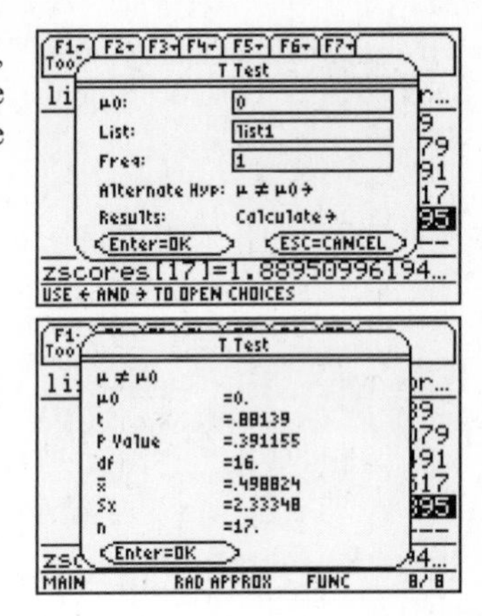

The results indicate there was no advantage from either lane. With a t statistic of 0.88 and a p-value of 0.3912, we fail to reject the null hypothesis of no difference. Our data do not indicate any unfairness due to lane assignment.

WHAT CAN GO WRONG?

Not Pairing Paired Data

This is a critical mistake. One needs to think carefully if there is some natural pairing of data that might (possibly) come from independent samples. Clearly, if the samples sizes are not the same, the data cannot have been paired. If one fails to pair data that should be paired, wrong conclusions will usually be made, due to overwhelming variability between the subjects. Here, we have the output if we had (wrongly) not used the paired test on the mileage data. Notice we would have made the opposite conclusion – the large p-value of more than 20% would indicate no difference in mileage due to shortening the work week.

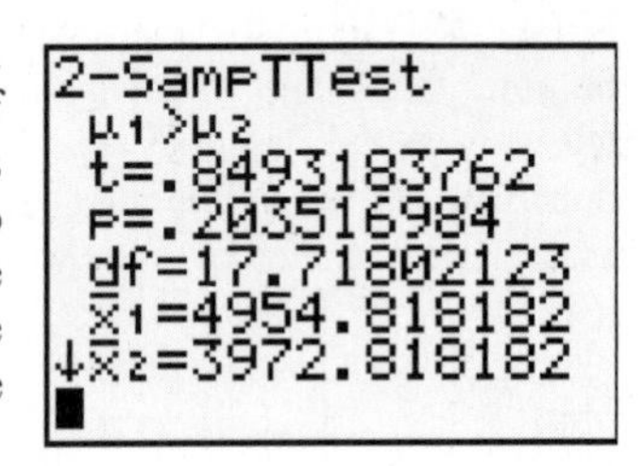

Bad Conclusions

The biggest thing to guard against is bad conclusions. Think about the data and what they show. Do not let conclusions contradict a decision to reject (this means we believe the alternate is true) or not reject (this means we have failed to show the null is wrong) a null hypothesis.

Other than that, there is not much that hasn't already been discussed – trying to subtract lists of differing length will give a dimension mismatch error. Having more plots "turned on" than are needed can also cause errors.

Chapter 12 – Comparing Counts

Count data are analyzed primarily for three different purposes: whether or not data agree with a specified distribution (a goodness-of-fit test), whether or not observed distributions collected at different times or places are consistent with one another (a test of homogeneity), and whether or not data classified according to two categorical variables indicate the categorical variables are related or not (a test of independence). All of these tests use a probability distribution called the $\chi 2$ (chi-squared) distribution. The first test, goodness-of-fit, is not a built-in function on the TI-83, but can be done easily enough. This is a built-in function with the latest operating system version on the TI-84 and on the 89 as well. The other two tests are built into all calculators and the mechanics are exactly the same whether testing homogeneity or independence; what is different is the setting and conclusions which can be made.

The $\chi 2$ statistic is defined to be $\sum \frac{(Obs - Exp)^2}{Exp}$ where the sum is taken over all the cells in a table. The quantities $\frac{(Obs - Exp)}{\sqrt{Exp}}$ are the standardized residuals which are examined in the event the null hypothesis is rejected to determine which cells deviated most from what is expected.

TESTING GOODNESS OF FIT

Does your zodiac sign determine how successful you will be in later life? Fortune magazine collected the zodiac signs of 256 heads of the largest 400 companies. Here are the number of births for each sign.

Births	Sign
23	Aries
20	Taurus
18	Gemini
23	Cancer
20	Leo
19	Virgo
18	Libra
21	Scorpio
19	Sagittarius
22	Capricorn
24	Aquarius
29	Pisces

We can see some variation in the number of births per sign- Pisces had the most and Gemini the least, but is it enough to claim that successful people are more likely to be born under certain star signs than others? If there is no difference between the signs, each should have (roughly) 1/12 of the births. That's the null hypothesis in this situation: births are evenly distributed across the year. The alternate is that the null is wrong: births are not evenly distributed across the year. These executives were a convenience sample, but we have no reason to suspect there was bias in their selection. Further, executives should be independent of one another and each category has expected frequency (1/12)*256 = 21.333 > 5. We may proceed with our inference.

TI-83 Calculations

```
L1      L2      L3    2
18      .08333
21      .08333
19      .08333
22      .08333
24      .08333
29      .08333
------
L2(13) =
```

Here, the observed counts have been entered into L1 and the (hypothesized) probability for each birth sign (1/12) has been entered into L2. The calculator displays the decimal equivalent for the entered fraction.
Note: Make sure both lists are the same length!

This next step calculates the quantities to be summed into the χ2 statistic. The expected values for each cell are the proportions times the total number of individuals in the study (or 256 in this example). One needs to be careful in entering the command – the parentheses are necessary! The general form of the command is (Obslist – n*Problist)^2/(n*Problist)[STO▸] Newlist.
Here the command was entered by pressing the following:
[(][2nd][1][−][2][5][6][×][2nd][2][)][x^2][÷][(][2][5][6][×][2nd][2][)][STO▸][2nd][3].
After pressing [ENTER] the first few entries are displayed.

```
(L1-256*L2)²/(25
6*L2)→L3
{.1302083333 .0…
```

We need to add the entries in L3 to find the χ2 value. Press [2nd][STAT] (List), arrow to MATH then press [5]to select option 5:sum(. The command shell is transferred to the home screen. Press [2nd][3] for L3 followed by [ENTER]. The χ2 statistic is 5.09375.

```
(L1-256*L2)²/(25
6*L2)→L3
{.1302083333 .0…
sum(L3
         5.09375
```

Now we need to get a p-value for the test. These tests are always one-tailed, so the p-value corresponds to area between 5.09375 and ∞ under a curve with k - 1 degrees of freedom, where k is the number of categories (here there are 12 categories, so 11 degrees of freedom). Press [2nd][VARS](DISTR). We want choice 7:χ2cdf(. Either arrow to the selection and press [ENTER] or press [7]. The command shell is transferred to the home screen.

The parameters are low end (5.09375), high end (properly [1][2nd][,][9][9] for infinity, but practically, several 9's will work), then the degrees of freedom. Be sure to separate the parameters with a comma. The p-value for the test is 0.927 which is very large. We will not reject the null hypothesis and conclude that these data indicate no significant birth sign differences among the executives.

```
6*L2)→L3
{.1302083333 .0…
sum(L3
         5.09375
χ²cdf(5.09375,99
9999,11)
    .9265413914
```

If the null were rejected, an examination of standardized residuals would show which cells were most different. The standardized residuals are just the square root of the entries in our list which had the components of χ2. To find them, from the home screen press [2nd][x^2][2nd][3][)][STO▸][2nd][4] followed by [ENTER]. To further examine these, use the Statistics list editor. We see (not surprisingly) the birth sign most different from its expected value is Pisces, since it was the largest entry in the list.

```
sum(L3
         5.09375
χ²cdf(5.09375,99
9999,11)
    .9265413914
√(L3)→L4
{.3608439182 .2…
```

TI-84/89 calculations

Using a TI-84 or -89 calculator, one can carry out the computations explicitly as indicated above for an 83, or use the built-in function. If you want to use the built-in function, the following section details its use. It is similar on both of these calculators.

The observed counts have been entered into L1 and the expected counts for each birth sign (1/12 of 256, or 256/12) has been entered into L2. The calculator displays the decimal equivalent for the entered fraction. If you are using a TI-89, enter the expected counts the same way.

L1	L2	L3 2
23	21.333	------
20	21.333	
18	21.333	
23	21.333	
20	21.333	
19	21.333	
18	21.333	

L2(1)=21.33333333…

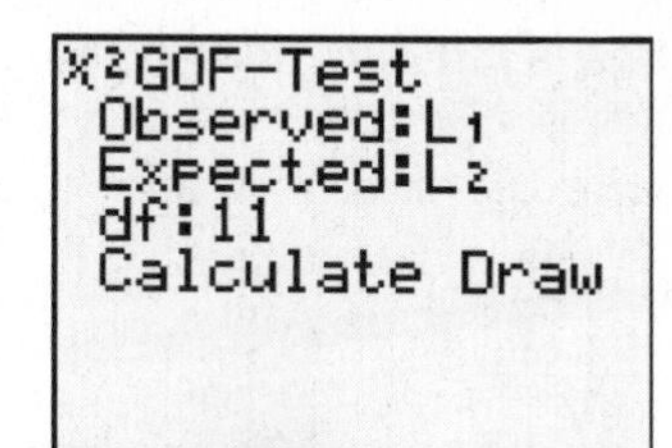

To perform the test, select `D:χ2GOF-Test` from the STAT, TESTS menu. Specify the two lists, and the degrees of freedom for the test, which are $k-1$ where k is the number of categories. Since we have 12 birth signs, the degrees of freedom are 11. Notice we have a choice of `Calculate` or `Draw`, just as with other hypothesis tests. `Calculate` gives more decimal places of output, and `Draw` highlights the area under the distribution curve which corresponds to the p-value of the test.

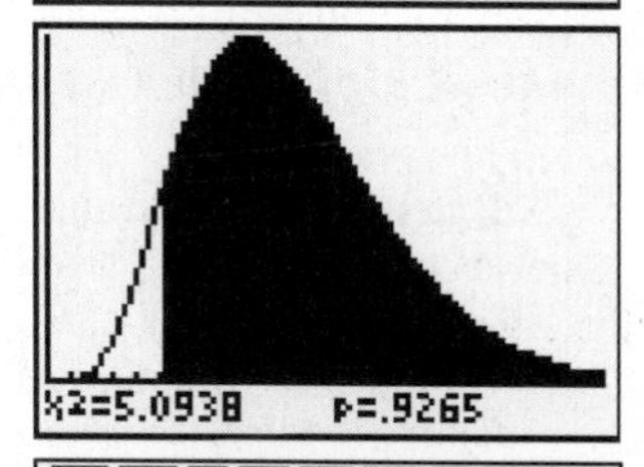

These are the results when `Draw` is selected. The calculated test statistic is $\chi 2 = 5.09$ and the p-value of the test is 0.9265. With these results, we fail to reject the null hypothesis and conclude, based on this data, the birth signs of executives are distributed evenly across the year – the observed differences were due to randomness.

If you are using a TI-89, the calculator automatically stores a new list of the components of the $\chi 2$ statistic. The square roots of these values are the standardized residuals (after appropriate positive and negative signs are attached). Scanning down this list we see (yet again) the birth sign most different from its expected value is Pisces, the last entry in the list.

TESTS OF HOMOGENEITY

Many colleges and universities survey graduating classes to determine their plans for the future. We might wonder whether plans depend on the student's major through their undergraduate college in a university. Here is a summary table from several universities for 2006 graduates. Each cell of the table shows how many students from each college (the columns) had that particular type of plan (the rows).

	Agriculture	Arts & Sciences	Engineering	Social Sciences	Total
Employed	379	305	243	125	1052
Grad School	186	238	202	96	722
Other	104	123	37	58	322
Total	669	666	482	279	2096

Visually, choices do not appear to be the same (look at the row for employment) but is the difference real or is it due perhaps to different size classes between the colleges? Since we have really the same distribution (future plans) for different colleges, this is a test of homogeneity: the null hypothesis is that the distribution of students' plans is the same across colleges, the alternate hypothesis is that the distributions are not the same.

Entering Data into a matrix: TI-83/84

We will first enter the numbers in the body of the table into a matrix. If you have a TI-83, press [MATRX]. If you have a TI-83+ or TI-84, press [2nd][x^{-1}]. Arrow to `EDIT`. Press [ENTER] to select matrix `A`.

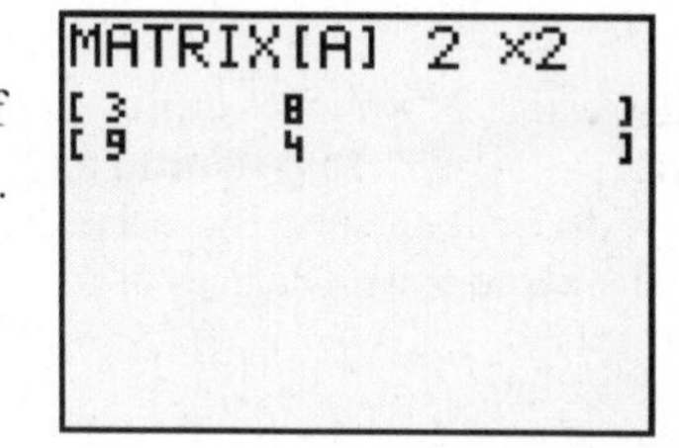

First we need to give the size of the matrix. The body of our table had three rows and four columns, so the matrix is 3 x 4. (We don't consider the column or row of totals as part of the data.) Press [4][ENTER][3][ENTER] to change the size of the matrix. Now type in the entries in the body of the table following each by [ENTER]. The process goes left to right, top to bottom.

Here is part of the filled-in matrix. Press [2nd][MODE](QUIT) to leave the Matrix Editor. Notice the small dashes at the right of the screen. These indicate there is more data "hidden" off to the right. To see it, press [▸].

Entering Data into a matrix: TI-89

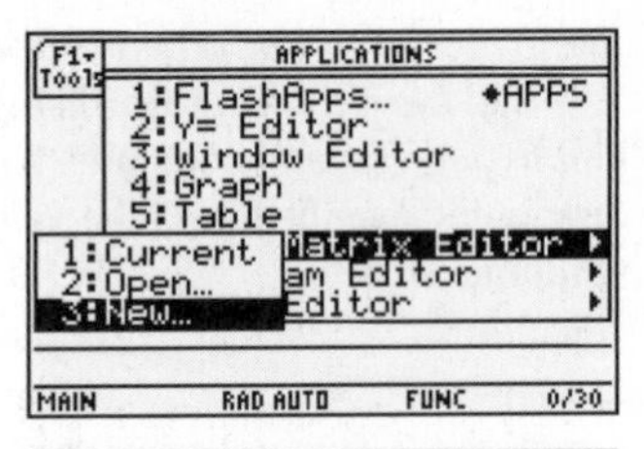

We will first enter the numbers in the body of the table into a matrix. From the home screen, press [APPS] Arrow to choice 6:Data/Matrix Editor and press the right arrow. Select choice 3:New. On the TI-89 Titanium, select the Data/Matrix Editor from the basic Applications screen.

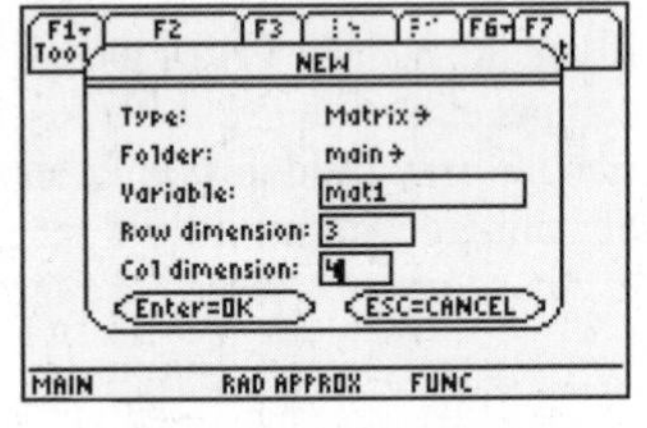

We first define the type. Press the right arrow and select Matrix. Leave the folder set as main unless you really want to change it. Press the down arrow and give the matrix a name. Here, to keep myself straight, I've named the matrix mat1. The cursor in this field is set to alpha by default, so pressing any number keys will result in their letter equivalent. If a number is desired, press [alpha] to change out of alpha mode. The body of our table had three rows and four columns, so the matrix is 3 x 4. Press [ENTER].

Cells are entered across rows. Press [ENTER] after each cell entry to proceed to the next. Here is part of the filled-in matrix. Press [2nd][ESC] (QUIT) to leave the Matrix Editor.

Performing the test

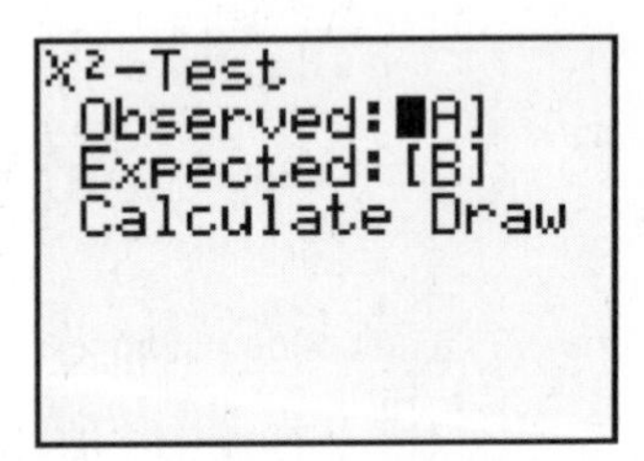

Now we're ready to perform the test. On a TI-83/84, press [STAT] then arrow to TESTS. We want choice C: χ2-Test. Either press [ALPHA][PRGM] (C) or press the up arrow to find the option followed by [ENTER]. All we need to tell the calculator is where the observed counts were entered and where to store expected counts. These default to matrix A for Observed and matrix B for Expected. If you need to change them, press [MATRX] or [2nd][x^{-1}] then select the desired matrix. Notice we again have choices Calculate and Draw.

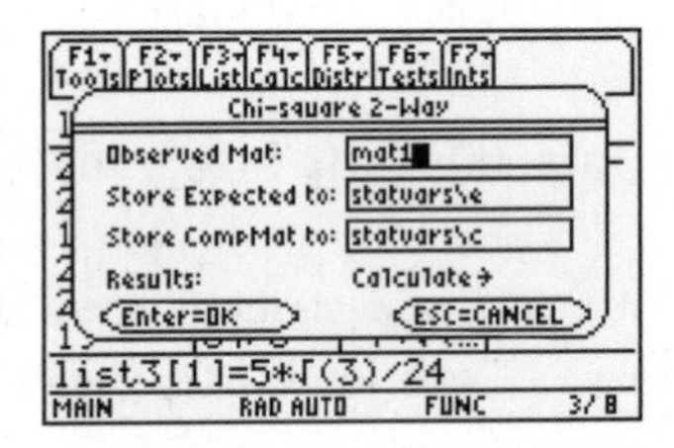

On an 89, return to the Statistics app and press [2nd][F1] for the Tests menu. Select choice 8:Chi2 2-way. To enter the name of the observed matrix, Go to the [VAR-LINK] screen ([2nd][−]) and find the name you gave it. The calculator will automatically store expected counts and components of the χ2 statistic in the named statistics matrices. Just as with the goodness-of-fit test, the (appropriately signed) square root of each component will be the standardized residual.

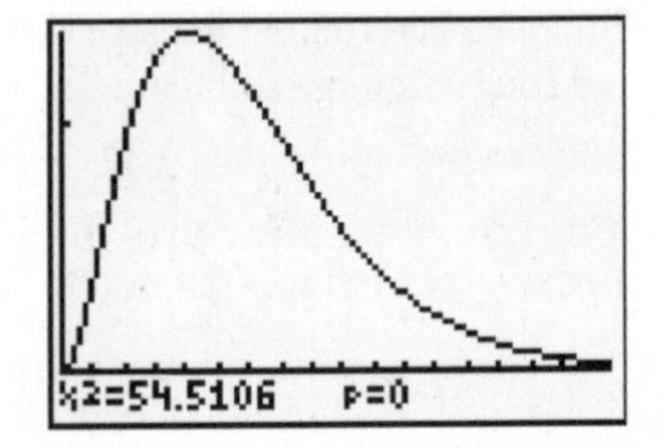

Selecting Draw gives the screen at right. We see three things here: the χ2 statistic (54.5106) and the p-value to four decimal places (0). We also see a different sort of distribution curve. χ2 curves are not symmetric; they are right skewed, and the shape varies with degrees of freedom (compare this curve to the one on page 71.)

If we had chosen `Calculate` instead of `Draw` we would have this screen. It shows the same statistic value, but a little more exact p-value (but this is still essentially zero). We are also given the degrees of freedom for the test which are $(rows-1)(cols-1)$, so (3-1)(4-1) = 6.

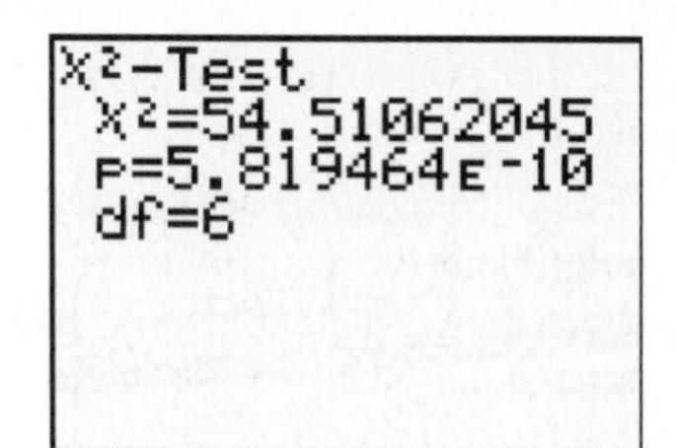

Since the p-value is so small, we reject the null hypothesis and conclude the distributions of post-graduation plans are not the same for all colleges.

Where are the differences? We'd like to get the standardized residuals. Unfortunately, the TI-83/84 won't give these easily. We can, however, compute the matrix Obs-Exp. The Observed counts are in matrix A and the Expected counts are in matrix B. From the home screen, press [MATRX], press [ENTER] to select matrix A, then [-] press [MATRX] press the down arrow and select matrix B by pressing [ENTER] then press [STO▶][MATRX] and select a new matrix (probably C). Lastly, another [ENTER] will perform the calculation and display some of the results.

Return to the matrix editor to see the entire matrix. The signs here indicate whether or not the observed count was more than the expected (positive) or less than expected (negative). Clearly the largest entries (in absolute value) are those for agriculture graduates. Many more of these than expected plan to be employed, while far fewer than expected have plans for graduate school.

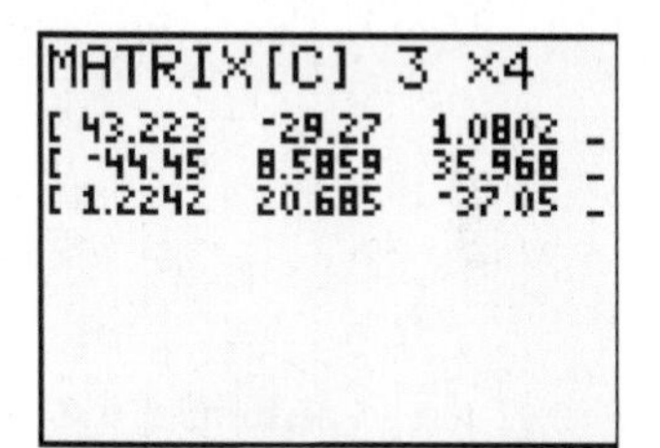

Here is matrix B which contains the expected counts. If we divide the entries in matrix C by the square root of the corresponding entry in matrix B we have the standardized residuals. The standardized residual agricultural graduates with employment plans is $43.223/\sqrt{335.78}=2.360$. Similarly we find the standardized residual for agriculture graduates with graduate school plans is -2.928.

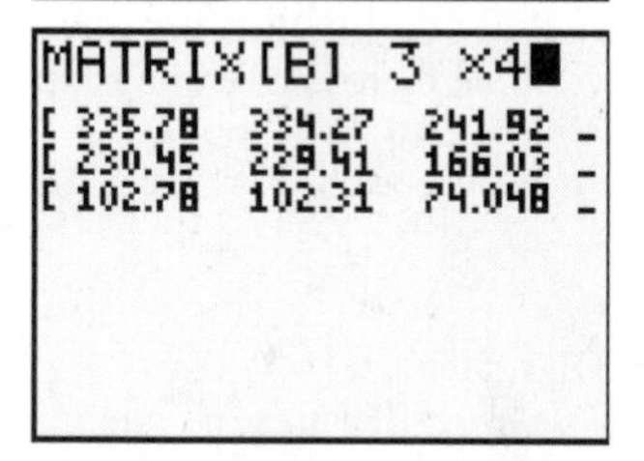

If you are using a TI-89, to see the components matrix, return to the Data/Matrix editor application, and select `Open`. On the next screen, you will be asked what type of data you want to open, the folder it is in, and the variable name. Here, I have indicated that I want to see a **`matrix`** in the **`statvars`** folder, namely, **`compmat`**, the components matrix. Use the right arrow to expand each selection, and the down arrow to change a value. Press [ENTER] to actually make the selection and proceed to the next entry. Pressing [ENTER] when done will display the matrix.

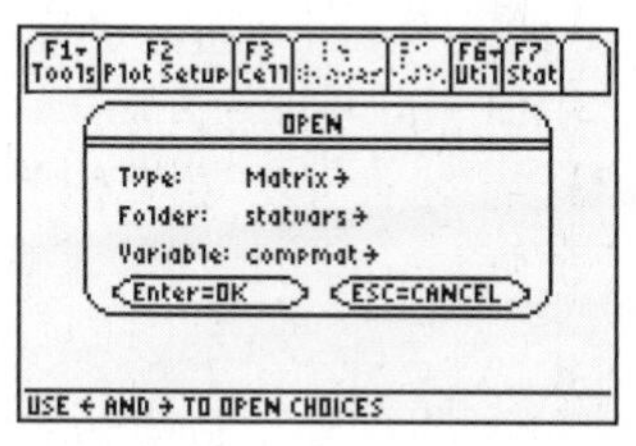

The components matrix shows the largest contribution to the χ2 statistic is in the third column, third row. This was the entry corresponding to engineering graduates with "other" plans (many fewer than expected). The next largest value was for agriculture graduates intending to pursue graduate school (again many fewer than expected). It seems that plans of university graduates are highly dependent on their field of study.

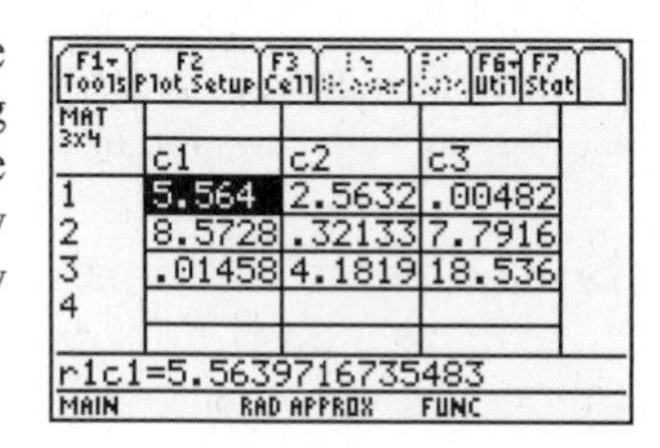

TESTING INDEPENDENCE

Tests of independence are used when the same individuals are classified according to two categorical variables. A study from the University of Texas Southwestern Medical Center examined whether the risk of Hepatitis C was affected by whether people had tattoos and by where they got their tattoos. The data from this study can be summarized in a two-way table as follows.

	Hepatitis C	**No Hepatitis C**
Tattoo, Parlor	17	35
Tattoo, Elsewhere	8	53
No Tattoo	22	491

Is the chance of having hepatitis C independent of (not related to) tattoo status? Our null hypothesis is that the two are not related. The alternate hypothesis is that there is a relationship.

Enter the numbers from the body of the table into a 3 x 2 matrix as discussed above. Press [2nd][MODE] (Quit) to exit the matrix editor.

The $\chi 2$ test is performed just as it is for the test of homogeneity. Here are the results. The p-value is extremely small. We will reject the null hypothesis and conclude there is a relationship between hepatitis C status and tattoos.

Here are the results of computing Obs-Exp and storing the result into a new matrix as above. The largest standardized residual is $13.096/\sqrt{3.9042} = 6.628$. People with tattoos from tattoo parlors are more likely than normal to have hepatitis C. Could it be that tattoo parlors are a source of hepatitis C?

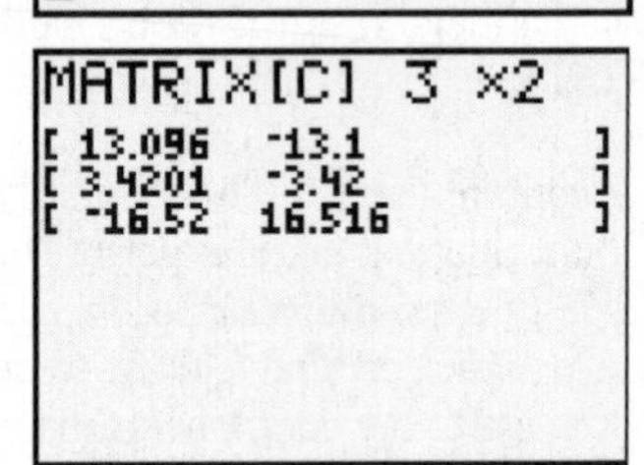

Here is the matrix of expected cell counts. Not all of them are more than 5. This means the conclusions based on the $\chi 2$ test we just performed may not be valid. Since the largest standardized residual is for one of these cells, this is a problem. A common solution is to combine cells in some manner to overcome the problem. In this case, both rows for tattooed people could be combined into one.

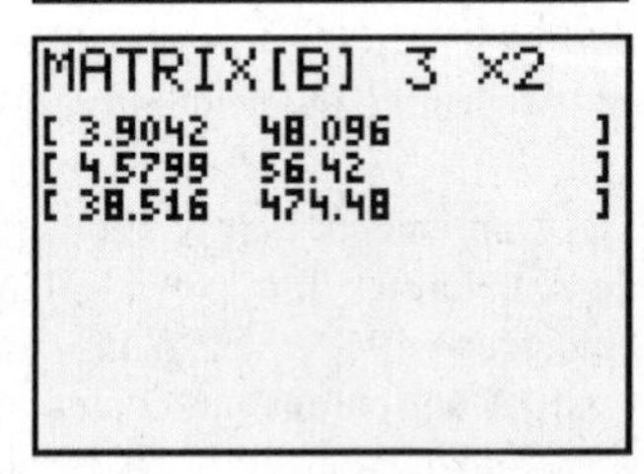

Race and Traffic Stops

Are Black drivers stopped more than others? Data from traffic stops in Cincinnati reported the race of the driver and whether the stop resulted in a search of the vehicle.[1] Is race a factor in vehicle searches? Race was recorded as Black, White, or Other. Since people who were stopped were classified both by race and whether or not they were searched, this is a test of independence.

[1] Police Vehicle Stops in Cincinnati, John E. Eck, Lin Liu, Growette Bostaph, Oct. 1, 2003, available at www.cincinnati-oh.gov.

	Black	White	Other	Total
Not searched	787	594	27	1408
Searched	813	293	19	1125
Total	1600	887	46	2533

This matrix is 2 x 3 because we have two rows (searched or not) and three columns for race.

MATRIX[A] 2 ×3
[787 594 27]
[813 293 19]
2,3=19

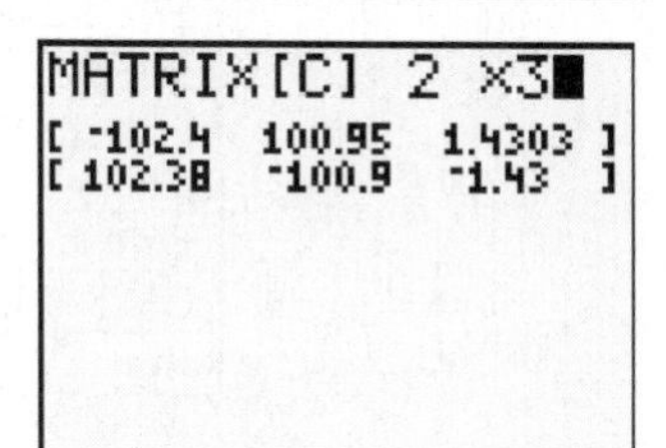

The $\chi 2$ statistic is 73.25 with a p-value 1×10^{-16} (essentially 0.) We have extremely strong evidence that whether or not a car was searched depends on the race of the driver. How are the variables related?

It is clear from looking at the differences Obs – Exp that Black drivers are more likely to be searched than white drivers. These are not the residuals (divide the entries in the matrix by the square root of the expected counts) but the direction of the relationship is clear.

MATRIX[C] 2 ×3
[-102.4 100.95 1.4303]
[102.38 -100.9 -1.43]

WHAT CAN GO WRONG?

Expected cell counts less than 5.

Check the computed matrix of expected cell counts. If they are not all greater than 5 the analysis may be invalid.

Missing or misplaced parentheses.

When computing elements for the goodness-of-fit test the parentheses are crucial.

Overusing the test.

These tests are so easy to do and data from surveys and such are commonly analyzed this way. The problem that arises here is that in this situation the temptation is to check many questions to see if relationships exist; but performing many tests on *dependent* data (the answers came from the same individuals) such as this is dangerous. In addition, remember that, just by random sampling, when dealing at $\alpha = 5\%$ we'll expect to see something "significant" 5% of the time when it really isn't. This danger is magnified when using repeated tests — it's called the problem of multiple comparisons.

Chapter 13 – Inference for Regression

Computing a regression equation and looking at residuals plots is not the end of the story. We might want to know if the slope (or correlation) is meaningfully different from 0. It's not always apparent that a slope is meaningfully non-zero. Consider these two equations for the selling price of a house: $price = 25 + 0.061 * sqft$ and $price = 25000 + 61 * sqft$. At first blush one might look at the small value for the slope in the first equation and believe it's reasonable to say the true slope may in fact be 0; however the difference is in the units – the first has price measured in thousands of dollars, the second in dollars. They're really the same line. In addition, we'd like to (perhaps) make a confidence interval for a "true" slope just as we did for means and proportions as well as confidence intervals for the average value of *y* for a given *x* and prediction intervals for a new *y* observation for an *x* value. TI-83/84 calculators can perform the t-test on the slope as a native function. The other functions can be performed either using the calculator output and tables for the *t*-distribution or with a program which is included in this manual. TI-84 calculators have a built-in function to compute confidence intervals for the slope; the TI-89 can do those as well as confidence intervals for a response

Returning to a problem considered before, here are advertised horsepower ratings and expected gas mileage for several model year 2007 vehicles.

Audi A4	200 hp	32 mpg	Honda Accord	166	34
BMW 328	230	30	Hyundai Elantra	138	36
Buick LaCrosse	200	30	Lexus IS 350	306	28
Chevy Cobalt	148	32	Lincoln Navigator	300	18
Chevy Trailblazer	291	22	Mazda Tribute	212	25
Ford Expedition	300	20	Toyota Camry	158	34
GMC Yukon	295	21	VW Beetle	150	30
Honda Civic	140	40			

How is horsepower related to gas mileage? Recall the plot that was constructed for this data in Chapter 5. It is reproduced at right. The trend is decreasing. The residuals plots in Chapter 5 showed no overt pattern against *X* (horsepower) and the normal probability plot was reasonably straight. Inference for the regression is therefore appropriate.

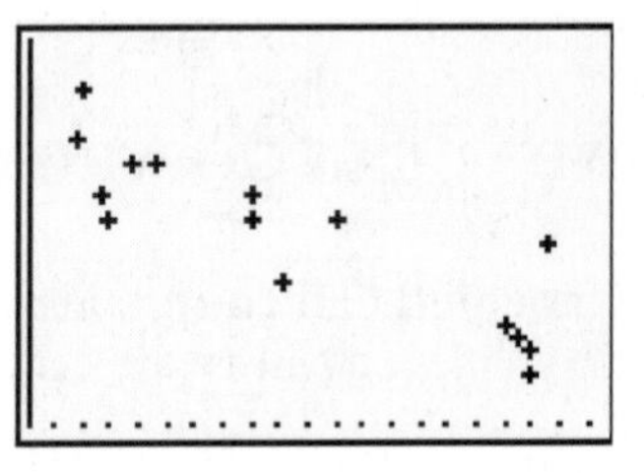

Press STAT, arrow to **TESTS** and select choice **E:LinRegTTest**. You tell the calculator which list contains the *x* (predictor variable) values, which contains the *y* (response) values. **Freq** is normally set to 1. Indicate the appropriate form of the alternate hypothesis. Notice there is an option to store the equation of the line. To store the equation as a function (here, Y_1), press VARS, arrow to **Y-VARS**, press ENTER to select **Function**, and ENTER to select Y_1. Finally, with the highlight on **Calculate**, press ENTER. The TI-89 input screen is similar, but there is an additional **Draw** option which, as usual shades the area under the *t* distribution corresponding to the p-value of the test.

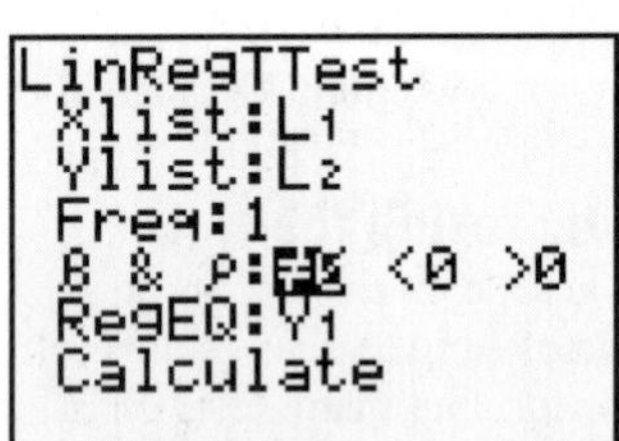

This is the first portion of the output (notice the ↓ at the bottom left). The first lines indicate the form of the regression so that you are reminded which quantity is the slope (b) and which the intercept (a) and the form of the alternate hypothesis in the test. The computed *t*-statistic for this regression is –6.32 and the p-value for the test is 0.00001, with 13 degrees of freedom. We will reject the null hypothesis and conclude not only that the slope is not zero; it is significantly negative. The intercept for the regression is 46.87.

```
LinRegTTest
 y=a+bx
 β<0 and ρ<0
 t=-6.322722868
 p=1.3237239E-5
 df=13
↓a=46.86797521
```

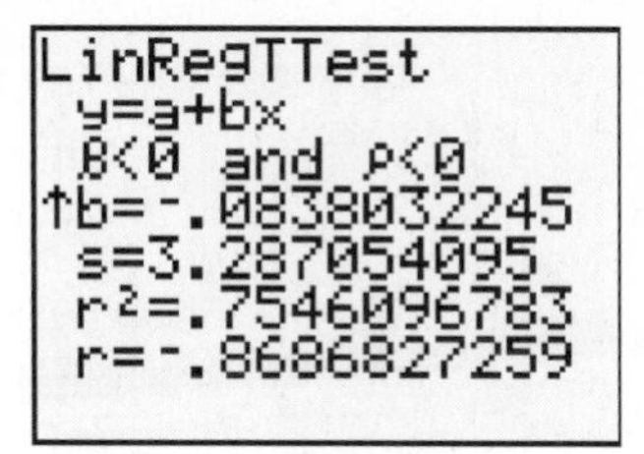

Pressing the down arrow several times we find the rest of the output. The slope is –0.084. We already know this is significantly different from zero even though its value seems small. The standard deviation of the data points around the line is 3.29. The relationship is strongly negative since $r^2 = 75.5\%$ and $r = -0.869$.

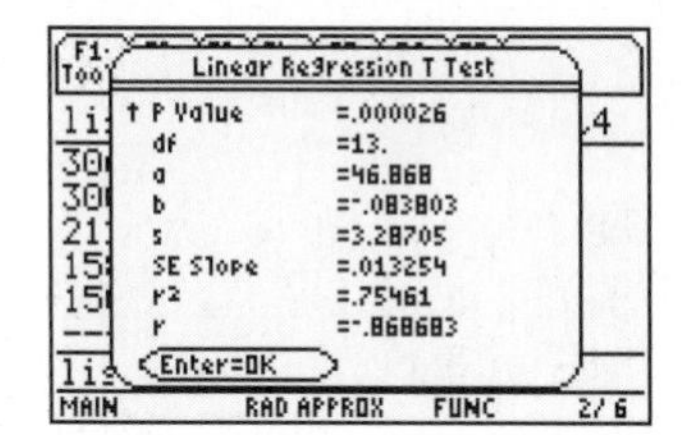

The output on a TI-89 includes all the quantities described above. It also gives the standard error of the slope (used in computing confidence intervals).

A CONFIDENCE INTERVAL FOR THE SLOPE

```
1-Var Stats
x̄=215.6
Σx=3234
Σx²=758754
Sx=66.28057246
σx=64.03311643
↓n=15
```

Confidence intervals (for any quantity) are always $estimate \pm (critical value)(SE(estimate))$. In this case the critical value of interest will be a t statistic based on 12 degrees of freedom. From tables, we find this is 2.179 for 95% confidence. The standard error of the slope is

$$SE(b_1) = \frac{s(e)}{\sqrt{\sum (x - \bar{x})^2}} = \frac{s(e)}{\sqrt{n-1} * s(x)}.$$

Using `1-Var Stats` for our horsepower data in `L1`, we find Sx is 66.281. We have all the pieces we need. $SE(b_1) = \frac{3.29}{\sqrt{14} * 66.281} = 0.0133$. When computing this, be sure to enclose the denominator in parentheses and close the parentheses for the square root. Putting all the pieces together, the 95% confidence interval for the slope is $-0.084 \pm 2.160 * 0.0133$ or (-0.113, -0.055). Based on this regression, we are 95% confident average gas mileage decreases between 0.055 and 0.113 miles per gallon for each horsepower in the engine.

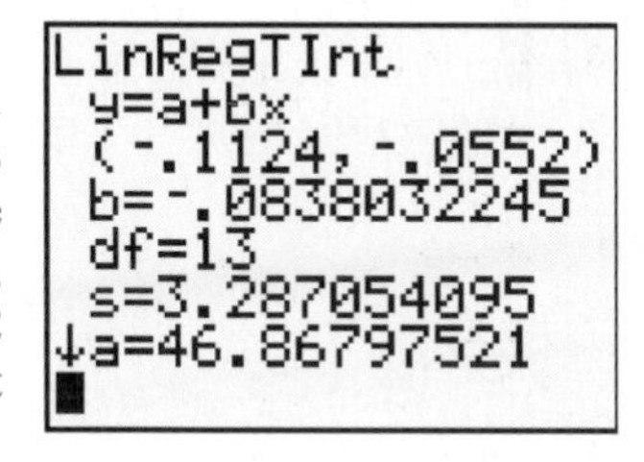

The TI-84 and -89 calculators can automatically compute the confidence interval for the slope. From the `STAT TESTS` menu, select `G:LinRegTInt`. This is option 7 on the `Ints` menu on the TI-89. The input screen is the same as for the t-test, except that it asks for the amount of confidence. From the output at right, we see we are 95% confident that gas mileage decreases between 0.055 and 0.112 miles per gallon for each horsepower in the engine. Notice this result is just slightly different from that obtained above due to rounding.

A CONFIDENCE INTERVAL FOR THE MEAN AT SOME *X*

What should we predict as the average gas mileage for a vehicle with 160 horsepower? Evaluating the equation for 160 horsepower gives 33.46 miles per gallon. This is just a point estimate, however and is subject to uncertainty just as any mean is. Confidence intervals account for this uncertainty. In this case there are two sources – average variation around the line as well as uncertainty about the slope which makes estimation more "fuzzy" further away from the mean. Both of these are accounted for in the equation of the standard error,

$$SE(\hat{\mu}_v) = \sqrt{s^2(b_1) * (x_v - \bar{x})^2 + \frac{s^2(e)}{n}}.$$

Putting everything together, we find $SE(\hat{\mu}_v) = \sqrt{0.0133^2 * (160 - 215.6)^2 + \frac{3.29^2}{15}} = 1.126$. The t critical value is still 2.16, so the confidence interval is 33.46± 2.16*1.126 or (31.03, 35.89). Based on this regression, we estimate with 95% confidence the average gas mileage for vehicles with 160 horsepower will be between 31.03 and 35.89 miles per gallon.

Using a TI-89, option `7:LinRegTInt` on the `Ints` menu will do all the calculations for us. Specify the lists containing the data, and select Response as the type of interval. Specify the x-value of interest, and the confidence level.

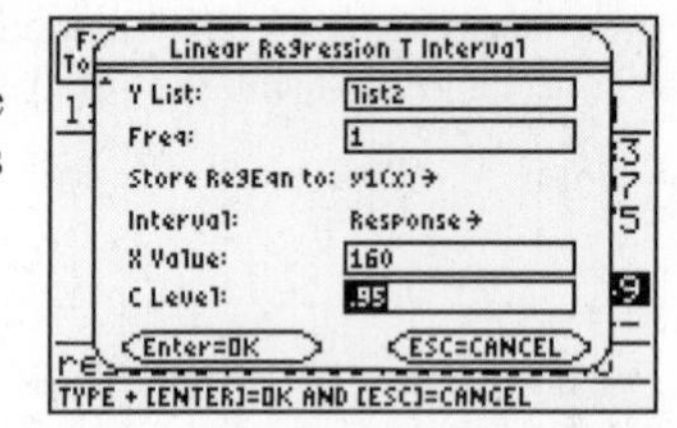

The first portion of output gives $\hat{y}$, the fitted value found by evaluation the equation at the x value of interest. We also see the confidence interval (31.03 to 35.89 miles per gallon is our 95% confidence estimate for average gas mileage for a vehicle with 160 horsepower, based on this sample), the margin of error for the estimate, the half width of the interval, and the standard error of the estimate.

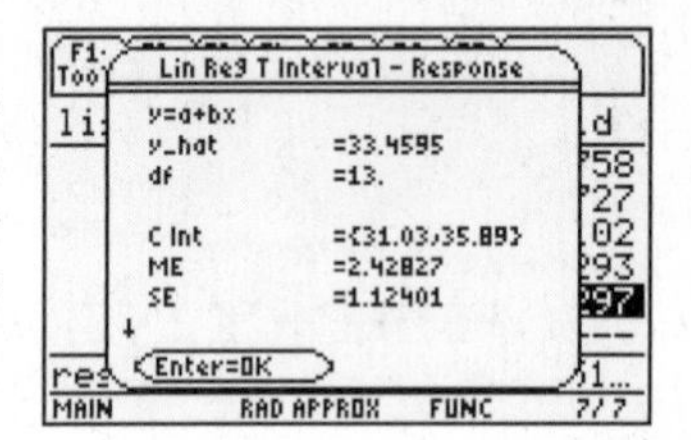

A PREDICTION INTERVAL FOR A NEW OBSERVATION

What would we predict for gas mileage for a particular vehicle with 160 horsepower? The point estimate is still 23.76 miles per gallon, but we have some additional uncertainly because individual observations are more variable than means. The standard error becomes $SE(\hat{y}_\nu) = \sqrt{s^2(b_1) * (x_\nu - \bar{x})^2 + \frac{s^2(e)}{n} + s^2(e)}$ which for our data becomes $SE(\hat{\mu}_\nu) = \sqrt{0.0133^2 * (160 - 215.6)^2 + \frac{3.29^2}{15} + 3.29^2} = 3.477$. So the prediction interval is $33.46 \pm 2.16 * 3.477$ or (25.95, 40.97). Based on this regression we estimate with 95% confidence the gas mileage for a vehicle with 160 horsepower will be between 25.95 and 40.97 miles per gallon.

With the TI-89, this interval is found by scrolling down the output from the `LinRegTInt` described above. We see the calculator found the same interval as we computed "by hand."

**A PROGRAM FOR REGRESSION INFERENCE

The author of this manual has written a program that performs these functions. (A listing is included and the program can also be obtained from the *Intro Stats* website which can be downloaded into a TI-83 or TI-84.) The program name is LSCINT. Once the program has been loaded into the calculator, to run the program, press PRGM and select that name from the list of programs. PgrmLSCINT is transferred to the home screen. Press ENTER to start the program.

You are prompted for the *X* list. enter its name and press ENTER. Do the same for the *Y* list.

```
prgmLSCINT
X LIST=L1
Y LIST=L2
```

The next screen gives the coefficients in the equation, the correlation coefficient (r) and the coefficient of determination, r^2 and well as the standard deviation of the residuals (s). Press ENTER to continue.

```
Y=a+bX
a=46.86797521
b=-.0838032245
r=-.8686827259
r²=.7546096783
S=3.287054095
```

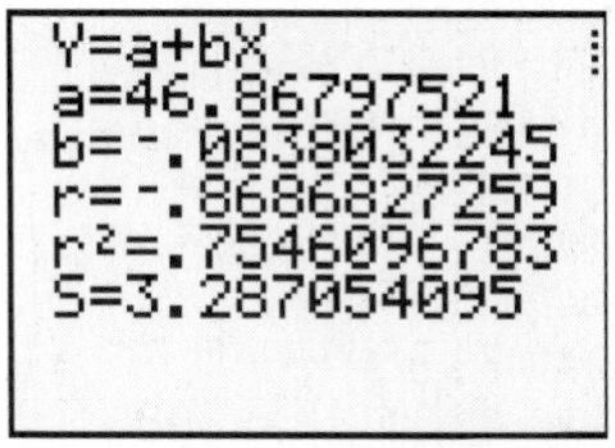

These are the results of the t-test for the slope. Press ENTER to continue.

```
t=-6.322722868
df=13
P=2.647447736E-5
```

You will next be prompted for a confidence level (enter it as a decimal) and the value of *X* for which confidence and prediction intervals will be created. Press ENTER after inputting each value.

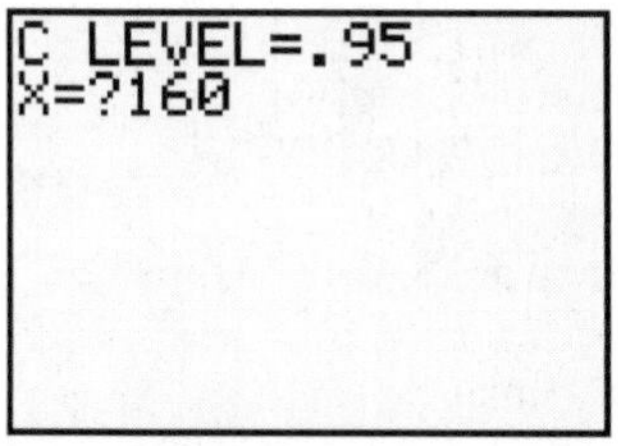

```
C LEVEL=.95
X=?160
```

The calculator displays the confidence interval for the slope. Press ENTER to continue.

```
C LEVEL=.95
X=?160
SLOPE CI=
 -.1124373847
 -.0551690644
```

Now the calculator displays the y-value for the given x and confidence and prediction intervals. Press ENTER to finish the program.

```
Y=33.45945928
MEAN CI
 31.03118796
 35.88773061
NEW Y CI
 25.954512
 40.96440656
```

WHAT CAN GO WRONG?

Assuming the lists are the same length, not much can go wrong that has not already been covered. One problem in doing many of these computations "by hand" comes from the compounding of round-off errors in intermediate computations. One is generally safest in using many digits in the interim and rounding only at the end. (Notice the "hand calculated" intervals are somewhat different from those obtained from the calculator. This is the reason.)

Program LSCINT listing (TI-83 Ascii version) This can also be found on the *Stats: Data and Modeling* website.

```
Input "X LIST=",LX
Input "Y LIST=",LY
FnOff
LinRegTTest LX,LY,0,Y1
σx²n[STO▸]V:s[STO▸]S:Ẍ[STO▸]M:n[STO▸]N
ClrHome
Output(1,2,"Y=a+bX"
Output(2,2,"a="
Output(2,4,a
Output(3,2,"b="
Output(3,4,b
Output(4,2,"r="
Output(4,4,r
Output(5,2,"r²="
Output(5,5,r²
Output(6,2,"S="
Output(6,4,S
Pause
df[STO▸]K
ClrHome
Output(1,1,"t="
Output(1,3,t
Output(2,1,"df="
Output(2,4,K
Output(3,1,"p="
Output(3,3,p
Pause
ClrHome
Input "C LEVEL=",C
Prompt X
K+1[STO▸]K:b[STO▸]A
TInterval 0, √ (K),K,C
upper[STO▸]T
A-T*S/√ (V)[STO▸]P
A+T*S/√ (V)[STO▸]Q
Output(3,1,"SLOPE CI="
Output(4,2,P
Output(5,2,Q
Pause
Y1-ST√(N⁻¹+(X-M)²/V)[STO▸]P
Y1+ST√(N⁻¹+(X-M)²/V)[STO▸]Q
Y1-ST√(1+N⁻¹+(X-M)²/V)[STO▸]J
Y1+ST√(1+N⁻¹+(X-M)²/V)[STO▸]U
ClrHome
Output(1,2,"Y="
Output(1,4,Y1
Output(2,1,"MEAN CI"
Output(3,2,P
Output(4,2,Q
Output(5,1,"NEW Y CI"
Output(6,2,J
Output(7,2,U
```

```
Pause
ClrHome
```

Chapter 14 – Analysis of Variance (ANOVA)

We have already seen two-sample tests for equality of the means in chapter 10. What if there are more than two groups? Answering the question relies on comparing variation among the groups to variation within the groups, hence the name. The null hypothesis for ANOVA is always that all groups have the same mean and the alternate is that at least one group has a mean different from the others.

Wild irises are beautiful flowers found throughout North America and northern Europe. Sir R. A. Fisher collected data on the sepal lengths in centimeters from random samples of three species. The data are below. Do these data indicate the mean sepal lengths are similar or different?

Iris setosa	Iris versicolor	Iris virginica
5.4	5.5	6.3
4.9	6.5	5.8
5.0	6.3	4.9
5.4	4.9	7.2
5.8	6.7	6.4
5.7	5.5	5.7
4.4	6.1	
	5.2	

TI-83/84 Procedure

I have entered the data into lists `L1`, `L2`, and `L3`. We will first construct side-by-side boxplots of the data for visual comparison. Visually, the medians are somewhat different with Iris virginica being the largest.

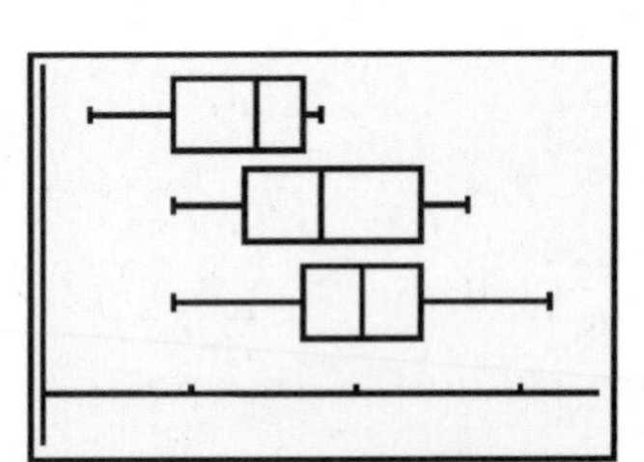

From the `STAT TESTS` menu select choice `F:ANOVA(`. This is the last test on the menu, so it is easiest to find by pressing the up arrow. The command shell is transferred to the home screen.

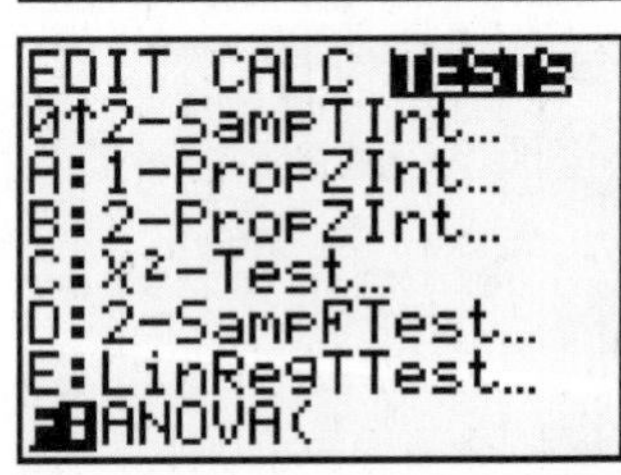

To complete the command, enter the list names separated by commas, then press [ENTER].

This is the first portion of the output. The value of the F statistic is 2.95 and the p-value for the test is 0.0779 which indicates at $\alpha = 5\%$ there is not a significant difference in the mean sepal lengths for the three species, based on this sample. The Factor degrees of freedom are k-1 where k is the number of groups, so with three groups, this is 2. MS is SS/df.

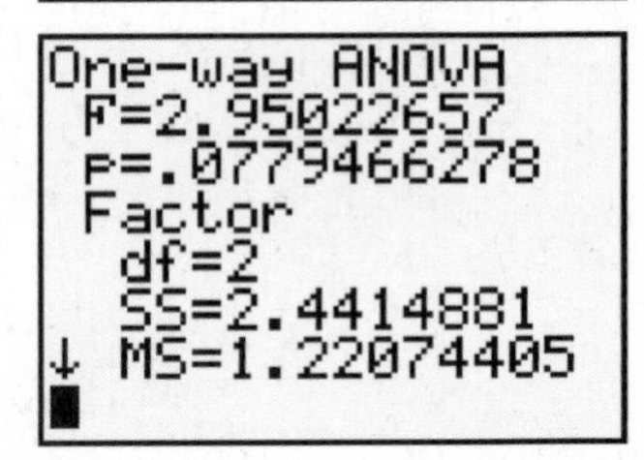

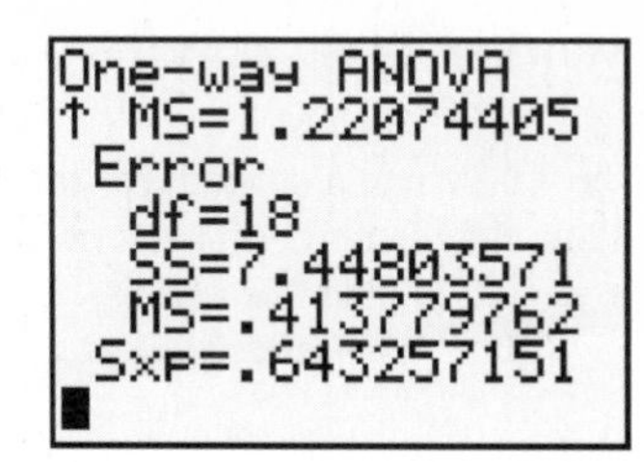

Pressing the down arrow several times gives the remainder of the output. Degrees of freedom for Error are $n-k$, where n is the total number of observations in all groups (21 here) and k is the number of groups (3). MS is again SS/df. Sxp is the estimate of the common standard deviation and is the square root of MSE. The F statistic is MSTR/MSE.

TI-89 Procedure

From the STAT Tests menu select choice C:ANOVA. This is the next-to-last test on the menu, so it is easiest to find by pressing the up arrow.

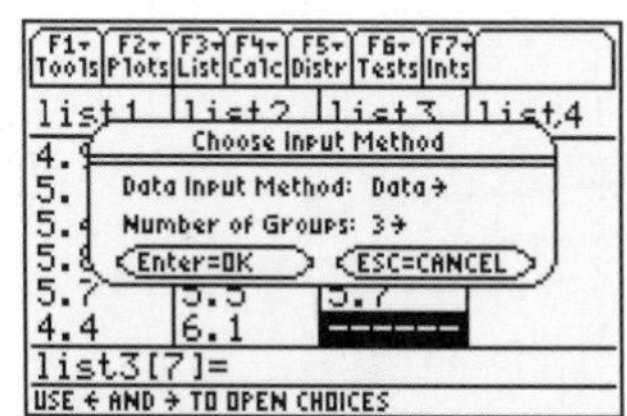

The first input screen asks if the data are in lists or if there are merely summary statistics. We also need to identify the number of groups (3 in this example). In each case, press the right arrow and make the appropriate selection. Press ENTER when finished to proceed.

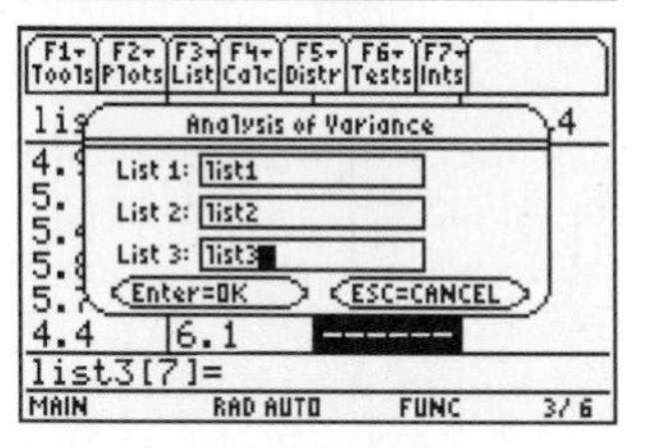

The second input screen asks for the list names. Use 2nd − (VAR-LINK) to access the list of list names and select those you have used.

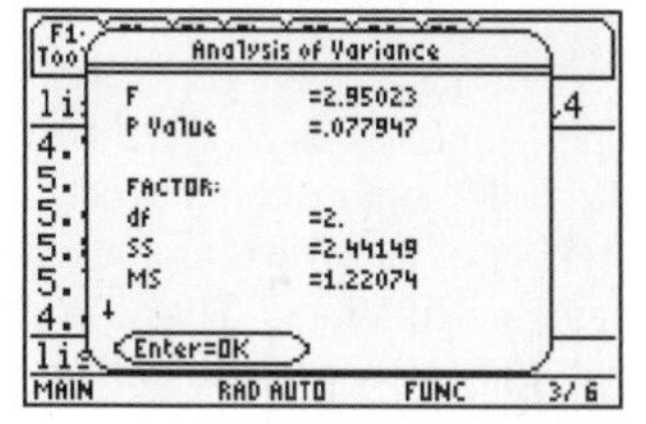

This is the first portion of the output. The value of the F statistic is 2.95 and the p-value for the test is 0.0779 which indicates at $\alpha = 5\%$ there is not a significant difference in the mean sepal lengths for the three species, based on this sample. The Factor degrees of freedom are k-1 where k is the number of groups, so with three groups, this is 2. MS is SS/df.

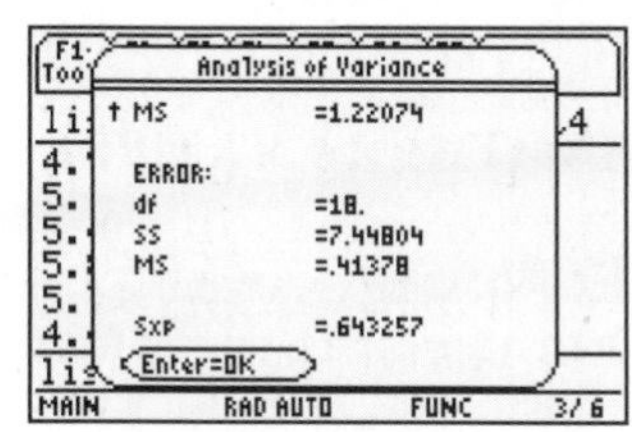

Pressing the down arrow several times gives the remainder of the output. Degrees of freedom for Error are n-k, where n is the total number of observations in all groups (21 here) and k is the number of groups (3). MS is again SS/df. Sxp is the estimate of the common standard deviation and is the square root of MSE. The F statistic is MSFactor/MSError.

list6	xbarl...	lowli...	uplist
------	5.2286	4.7178	5.7394
	5.8375	5.3597	6.3153
	6.05	5.4983	6.6017
	------	------	------

xbarlist[1]=5.22857142857...
MAIN RAD AUTO FUNC 7/9

After pressing ENTER to clear the output screen, we find something has been added into the Statistics editor. The first column contains the means for each group. The next two lists, lowlist and uplist, contain the lower and upper limits of individual 95% t-intervals for the mean, where Sxp has been used as the common standard deviation. Looking at these is a crude method of determining whether group means are equal or not. If the intervals have a large amount of overlap it is an indication that group means are not different. Since all of these intervals overlap, it is a confirmation of the decision that the mean sepal lengths of the three species are similar.

Another TI-89 example

The TI-89 can also perform the analysis using the summary statistics for each group. Your text uses an example about the number of bacteria left on hands after washing with different types of cleanser. The statistics are presented in the table below.

Level	n	Mean	St. Dev.
Alcohol Spray	8	37.5	25.56
Antibacterial soap	8	92.5	41.96
Soap	8	106.0	46.96
Water	8	117.0	31.13

On the first Anova screen, I have indicated that I am using the Stats method of input, and that there are four groups.

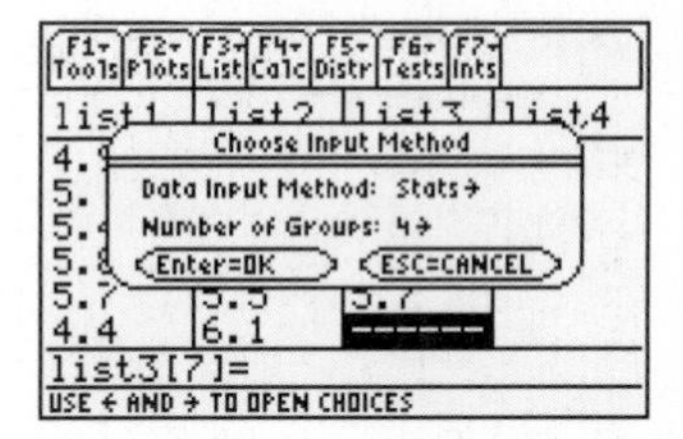

Input the summary statistics for each group in the order specified, and enclosed by curly braces. These are 2nd (and 2nd). Failure to include the curly braces will result in an error message that (at least) one is missing.

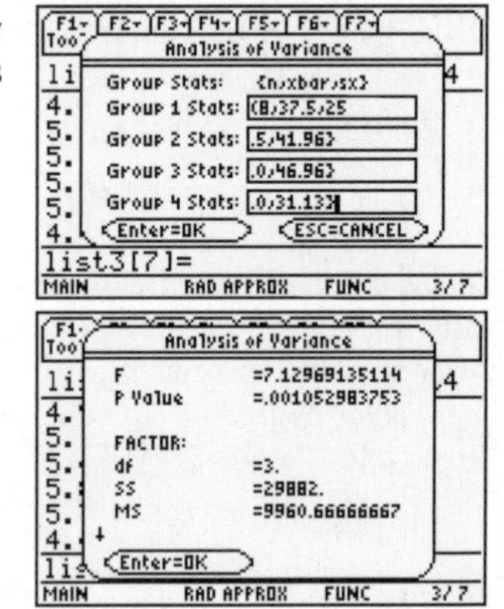

The output indicates gives the F statistic 7.13, with a p-value of 0.0011. We reject the null hypothesis and conclude that the method of hand washing does affect the number of bacteria left on the skin. Note that there is some rounding error present in my calculations. This is probably due to the fact that I used the (rounded) summary statistics, rather than the actual data.

ANOTHER EXAMPLE

The following data represent yield (in bushels) for plots of a given size under three different fertilizer treatments. Does it appear the type of fertilizer makes a difference in mean yield?

Type A	Type B	Type C
21	41	35
24	44	37
31	38	33
42	37	46
38	42	42
31	48	38
36	39	37
34	32	30

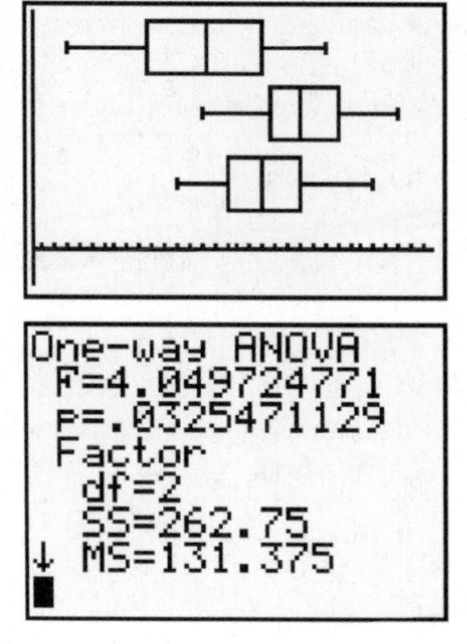

Here are side-by-side boxplots of the data. All three distributions appear symmetric and there is not a large difference in spread, so ANOVA is appropriate. Using `1-Var Stats` we find the mean for Type A is 32.125 bushels, Type B has a mean of 40.125 bushels, and Type C has a mean of 37.25. They're different, but are they different enough to believe the three types of fertilizer have different effects on crop yield?

The ANOVA output of interest is at right. We see the F-statistic is 4.05 and the p-value for the test is 0.0325. At $\alpha = 0.05$, we will reject the null hypothesis and conclude that at least one fertilizer has a different mean yield than the others. Scrolling down, we further find `Sxp` = 5.696.

WHICH MEAN(S) IS (ARE) DIFFERENT?

Having rejected the null hypothesis, we would like to know which mean (or means) are different from the rest. For a rough idea of the difference one can do confidence intervals for each mean (using the pooled standard deviation found in the ANOVA) and look for intervals which do not overlap, but this method is flawed. Similarly, testing each pair of means (doing three tests here) has the same problem: the problem of multiple comparisons. If we constructed three individual confidence intervals, the probability they all contain the true value is $0.95^3 = 0.857$, using the fact that the samples are independent of each other.

The Tukey Method

The Tukey method is one way to solve this question. From tables, we find the critical value q^* for k groups and $n-k$ degrees of freedom for error. My table gives $q^*_{3,21,0.05} = 3.58$. The "honestly significant difference" is computed as $HSD = \frac{q^*}{\sqrt{2}} s_p \sqrt{\frac{2}{n_i}}$ where each group has n_i observations. (NOTE: most tables of this distribution require the division of q^*, but not all. Check your particular table.) Here, $HSD = \frac{3.58}{\sqrt{2}} * 5.695\sqrt{\frac{2}{8}} = 7.208$. Groups with means that differ by more than 7.208 are significantly different. The mean for Type A was 32.125; for Type B 40.125 and for Type C 37.25. We therefore declare types A and C have similar means; Types B and C also have similar means. The means which are different from each other are those for types A and B.

The Bonferroni Method

The Bonferroni method is based on t distributions, unlike the Tukey method. One disadvantage is that you will require a calculator with the inverse t function, or a computer to find exact t^* multipliers. Another disadvantage with this method is that you must know ahead of time how many comparisons you will be doing.

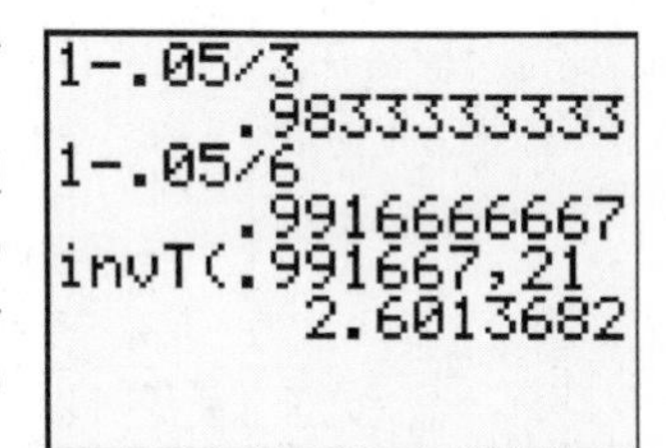

The Bonferroni method for making j comparisons, involves computing a "confidence interval" for each comparison using a confidence level of $1-\alpha/j$ which entails finding a critical value t^{**} with $n-k$ degrees of freedom, and $1-\alpha/2j$ tail probability. For our fertilizer data we want three comparisons, so for 95% overall confidence we need a 98.33% confidence interval, or t^{**} for 99.16% area below the critical value. My calculator indicates the critical value is 2.601.

Our data had 8 observations in each group, and a pooled standard deviation of 5.696, so the least significant difference with this method (the margin of error for a confidence interval for the difference) becomes

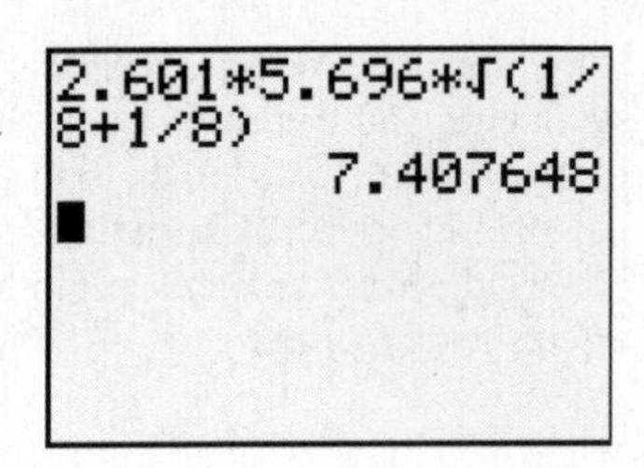

$LSD = t^{**} \times s_{xp}\sqrt{\frac{1}{n_1}+\frac{1}{n_2}} = 2.601*5.696\sqrt{\frac{1}{8}+\frac{1}{8}} = 7.408$. Any two group means that are farther apart than 7.408 will be called significantly different. Again, Types A and B are different.

Notice the Bonferroni least significant difference is slightly larger than the Tukey quantity. If you have many groups, and want comparisons for all possible pairs, the Bonferroni method will yield much larger LSDs than the Tukey method.

Chapter 15 – Multiple Regression

Multiple regression is an extension of the linear regression already studied where we create a model to explain a response variable based on more than one predictor. Just as with linear regression, we will want to examine how well the predictors determine the response, individually and as a group, by testing the utility of the model and create confidence intervals for slopes, mean response, and predictions of new responses. TI-83/84 calculators do not have a built-in multiple regression capability. TI-89 calculators do have this ability. If you are using a TI-83 or -84, the author of this manual has written a program (provided on the text's web site with a listing at the end of this chapter) called MULREG to do this.

How well do age and mileage determine the value of a used Corvette? The author chose a random sample of ten used Corvettes advertised on autos.msn.com. The data are below.

Age (Years)	Miles (1000s)	Price ($1000)
3	46	27
1	11	43
2	20	35.5
1	11.5	39
8	69	16.5
5	49	23
2	10	38
4	27	32
5	30.5	30
3	46	27

We first examine plots of each predictor variable against Price. The plot against age is linear, and decreasing as expected (we expect older cars to cost less). The regression equation for this relationship is *Carprice* = 42.44 – 3.33**Age*, with r^2 = 80.6%. This suggests the average price of a used Corvette goes down $3330 each year.

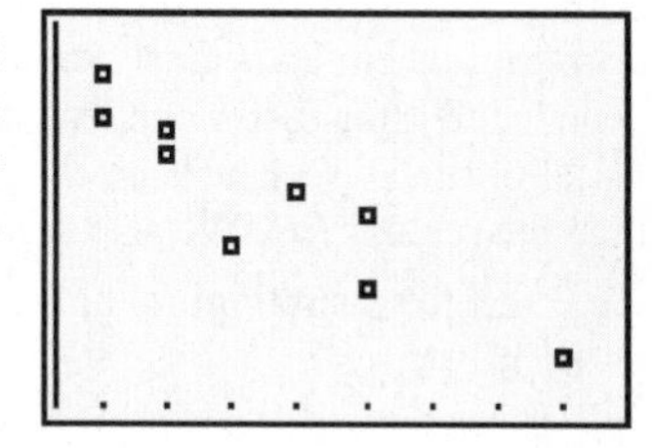

The plot of price against mileage is also linear and decreasing with even less scatter than in the other plot. The regression equation for this relationship was found to be *Carprice* = 43.77 – 0.40**Miles*, with r^2 = 95.5%. This suggests the average price of a used Corvette will decrease $400 for every 1000 miles it has been driven.

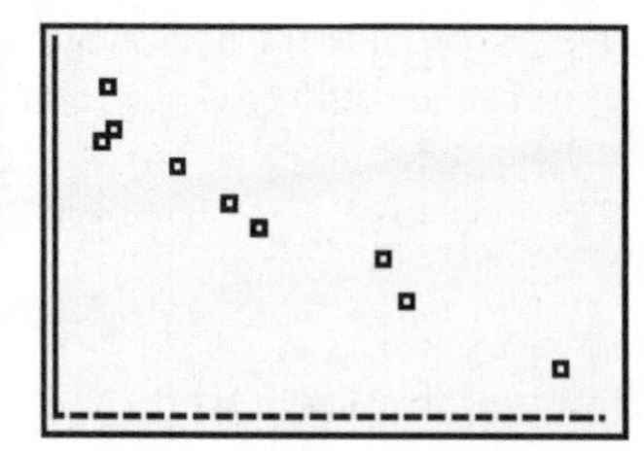

TI-83/84 Procedure

First the data are entered into Matrix [A]. On the TI-83, press [MATRX]; if you have a TI-83+ or TI-84, Matrix is [2nd][x^{-1}]. Arrow to EDIT and press [ENTER] to select matrix [A]. Enter the number of observations (or rows; here that is 10) and the number of columns (3). Type in the data, pressing [ENTER] after each number, across the rows of the matrix. Press [2nd][MODE] (Quit) to exit the editor.

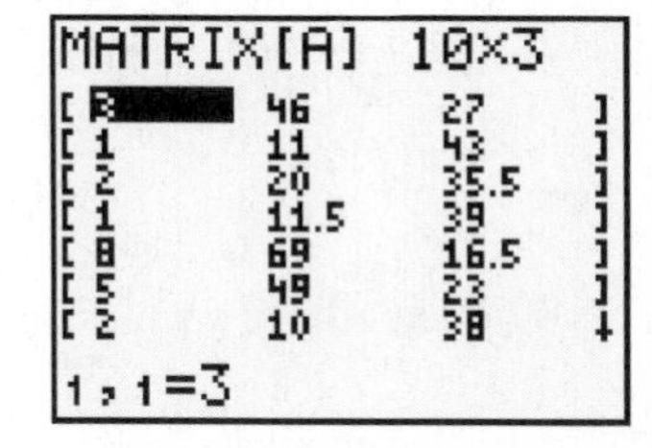

With the program transferred to the calculator, pressing [PRGM] will give the list of all programs stored. Select MULREG, then press [ENTER] to start it running. You will first be asked which column of the matrix has the response (Y) variable. In our example, price is in the third column. Press [ENTER] after the response.

```
                 Done
prgmMULREG
DATA IN COLS
OF [A]
RESPONSE COL=3
```

Here is the first portion of the output, the coefficients. These are displayed with the program paused so you can use the right arrow to scroll through them. They are also stored in list θ1 which can be accessed under the LIST menu ([2nd][STAT]). Press [ENTER] to resume execution of the program. We find the equation of the model is $Price = 44.200 - 0.943 * Age - 0.309 * KMiles$.

```
                 Done
prgmMULREG
DATA IN COLS
OF [A]
RESPONSE COL=3
COEF Lθ1=
{44.2002058 -.9…
```

The next quantities displayed are the standard deviations of the coefficient estimates. As with the coefficients themselves, the calculator is paused to allow scrolling through the list. These are stored in list θ2 for further use. After pressing [ENTER], the t statistics for each coefficient are displayed in a like manner, followed by the p-values for testing the hypotheses H_0: $\beta_i = 0$ against H_A: $\beta_i \neq 0$.

```
RESPONSE COL=3
COEF Lθ1=
{44.2002058 -.9…
STDEV Lθ2=
{.9457508901 .4…
T-RATIO Lθ3=
{46.7355688 -2.…
```

The next screen first displays S, the standard deviation of the residuals, then R^2, the coefficient of determination, 97.4%, which is how much of the variation in response (price) is explained by the model (in this case the age of the car and its mileage). The next quantity shown is the adjusted R^2. Since R^2 can never decrease when additional variables are added into a model, this quantity is "penalized" for additional variables which do not significantly help explain variation in Y, so it will go down if this is the case. Adjusted R^2 is always less than the regular R^2. We also see the degrees of freedom which are associated with the F-test for overall utility of the model. Press [ENTER] to resume execution of the program.

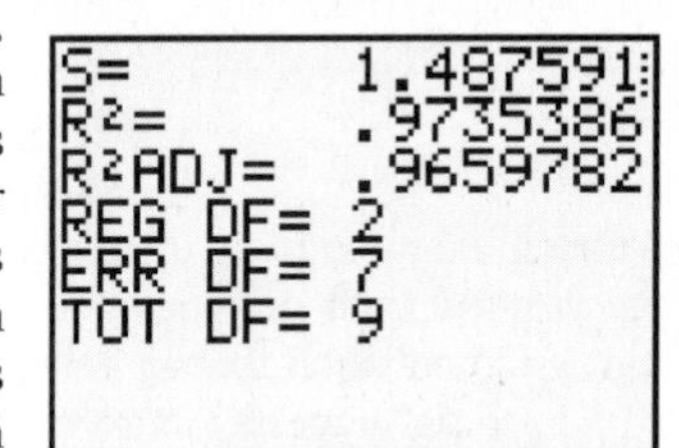

This screen shows the sums of squares and the F statistic for the overall significance (utility) of the model, along with its p-value. Here the p-value (to 5 decimal places) is 0 indicates there is a significant relationship between price and its two predictors.

```
SS REG= 569.9095
SS ERR= 15.49049
SS TOT= 585.4
MS REG= 284.9547
MS ERR= 2.212928
F=      128.7681
P-VAL=  0
```

Now we are asked if we want to use the model to predict a value based on the equation, or quit. Make the appropriate choice.

TI-89 Procedure

From the Statistics list editor, press [2nd][F1] (Tests). Select menu option B:MultRegTests.

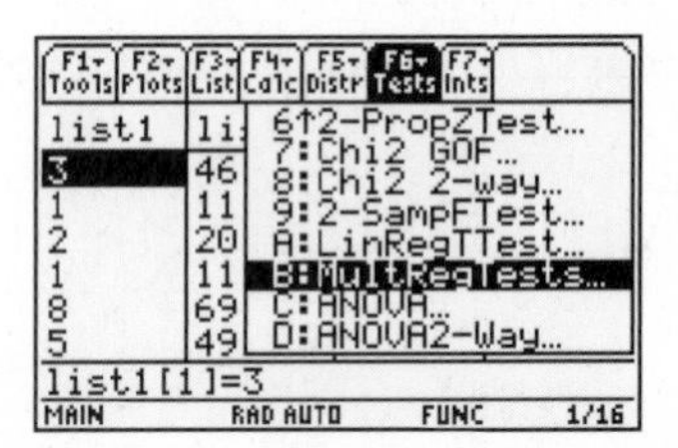

The input screen first asks how many independent variables there are. Use the right arrow to change this to the proper number. Enter the list names for the Y list and the X lists using [2nd][-] (VAR-LINK). Press [ENTER] to perform the calculations when all list names have been entered.

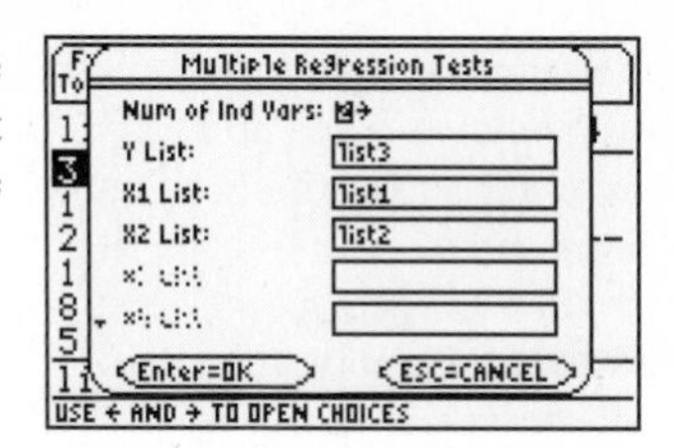

Here is the first portion of the output. The first line is the form of the regression equation. The second and third lines give the F statistic for the overall significance (utility) of the model, along with its p-value. Here the p-value of 0.000003 indicates there is a significant relationship between price and its two predictors. R^2 which is 97.3 % in this situation is the amount of variation in price which is explained by the two predictors (age and mileage).

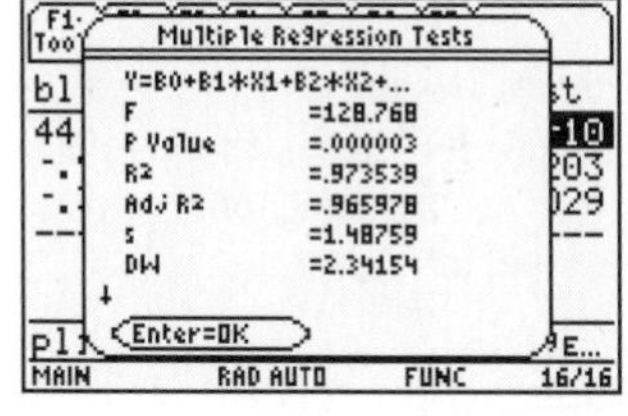

The next quantity is the adjusted R^2. Since R^2 can never decrease when additional variables are added into a model, this quantity is "penalized" for additional variables which do not significantly help explain variation in Y, so it will go down if this is the case. Adjusted R^2 is always less than the regular R^2. S is the standard deviation of the residuals.

DW is the value of the Durbin-Watson statistic which measures the amount of correlation in the residuals and is useful for data which are time series (data that have been collected through time). If the residuals are uncorrelated, this statistics will be about 2 (as it is here); if there is strong positive correlation in the residuals, DW will be close to 0; if the correlation is strongly negative, DW will be close to 4. Since these data are not a time series, DW is meaningless for our example.

Pressing the down arrow we find the components for regression and error which are used in computing the F statistic. The F statistic for regression is the MS(Reg)/MS(Error) where the Regression Mean square functions just like the treatment (factor) mean square in ANOVA.

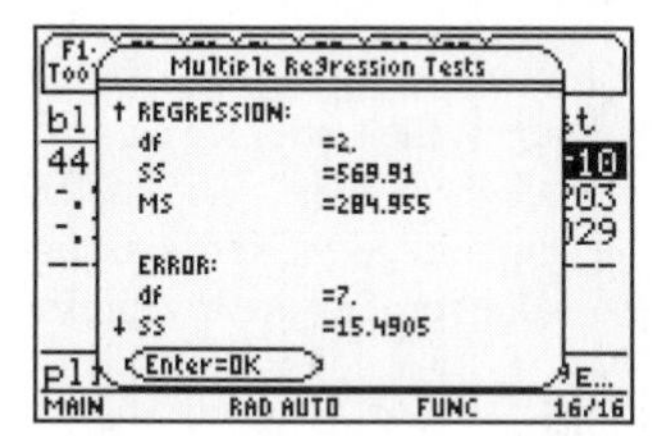

Finally we see some of the entries in new lists that have been created. The complete lists will be seen when [ENTER] is pressed. `Blist` contains the estimated intercept and coefficients; `SE list` is the list of standard errors for the coefficients which can be used to create confidence intervals for true slopes; `t list` gives values of the t-statistics for hypothesis tests about the slopes and intercept; `P list` gives the p-values for the tests of the hypotheses H_0: $\beta_i = 0$ against H_A: $\beta_i \neq 0$. If the assumed alternate is 1-tailed, divide these p-values by 2 to get the appropriate p-value for your test.

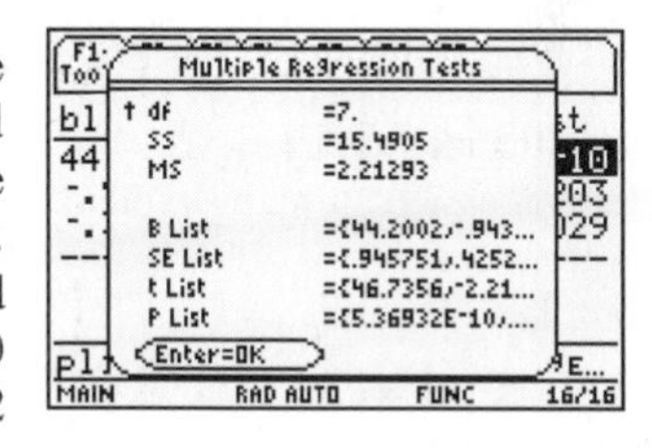

After pressing [ENTER] we see several new lists that have been added into the editor. `Yhatlist` is the list of predicted values for each observation in the dataset based on the model ($yhat_i = b_0 + b_1x_{1i} + b_2x_{2i}$ in this model); `resid` is the list of residuals $e_i = y_i - yhat_i$. `Sresid` is a list of standardized residuals obtained by dividing each one by S, since they have mean 0. If the normal model assumption for the residuals is valid, these will be N(0, 1).

zscor...	yhatl...	resid	sresid
------	27.149	-.1494	-.1283
	39.856	3.1438	2.4267
	36.131	-.6305	-.4591
	39.702	-.7016	-.5409
	15.322	1.178	1.2537
	24.335	-1.335	-.9918

yhatlist[1]=27.1493826141...

Pressing the right arrow we find yet more lists. Leverage is a measure of how influential the data point is. These values range from 0 to 1. The closer to 1, the more influential (more of an outlier in its *x* values) the point is in determining the slope and intercept of the fitted equation. Values greater than $2p/n$ where n is the number of data points and p is the number of parameters in the model are considered highly influential. Here, $n = 10$ and $p = 3$, so any value grater than 0.6 will designate an observation as highly influential. This indicates in our example that the point for the 8-year-old car is influential.

lever...	cookd	blist	selist
.38772	.00348	44.2	.94575
.24161	.62537	-.9434	.42529
.14756	.01216	-.3091	.04637
.23948	.0307	------	------
.601	.78915		
.18114	.07253		

leverage[1]=.387715535021...

Cook's Distance in the next column is another measure of the influence of a data point in terms of both its *x* and *y* values. Its value depends on both the size of the residual and the leverage. The ith case can be influential if it has a large residual and only moderate leverage, or has a large leverage value and a moderate residual, or both large residual and leverage.

To assess the relative magnitude of these values, one can compare them against critical values of an *F* distribution with p and $n - p$ degrees of freedom or use menu selection `A: F Cdf` from the [F5] (`Distr`) menu. The largest value in the list is for the Corvette data is again for the fifth observation (the 8-year-old). This is the input screen. We are finding the area above 0.78915 which is the largest value in the list. The result is 0.5372, which indicates this is not unusual, so this point is not influential.

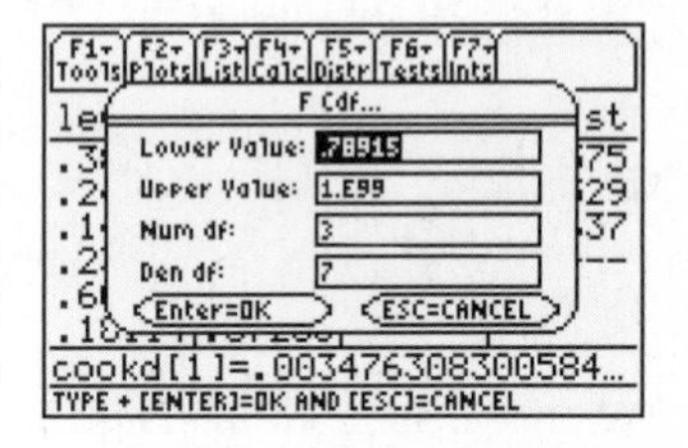

After pressing the right arrow still more, we find the last of the output lists. `Blist` is the list of coefficients. We finally see the fitted regression equation:
$Price = 44.2 - 0.94 * Age - 0.31 * Miles$. We interpret the coefficients in the following manner: price declines $940 on average for each year of age when mileage is the same; for cars of a given age, every additional 1000 miles reduces average price $310. The coefficient of miles is similar to that obtained from the simple regression ($400) but the decrease for age is much less than the value for the simple regression ($3300).

blist	selist	tlist	plist
44.2	.94575	46.736	5.E-10
-.9434	.42529	-2.218	.06203
-.3091	.04637	-6.667	.00029

plist[1]=5.3693177746919E...

The next column contains the standard errors of each coefficient. These can be used to create confidence intervals for the true values using critical values for the *t* distribution for $n - p$ degrees of freedom. Finally we see the *t* statistics and p-values for testing H_0: $\beta_i = 0$ against H_A: $\beta_i \neq 0$. These suggest the coefficient of age is not significantly different from 0; in other words, mileage is a much more determining quantity for the price of used Corvettes which helps explain why its coefficient changed less than the coefficient of age from the single variable regressions.

ASSESSING THE MODEL

Just as with simple (one-variable) linear regression, we will use residuals plots to assess the model. The program has stored fitted values (y-hats) in `L5` and residuals in `L6`.

Using the `STAT PLOT` menu, define a scatter plot of the residuals in `L6` (as Y) against the fitted values in `L5`. Pressing [ZOOM][9] displays the plot. Just as with simple regression we are looking in this plot for indications of curves or thickening/narrowing which indicate problems with the model. With this small data set these are somewhat hard to see, but clearly the only positive residuals are for the largest and smallest fitted values, which could indicate a potential problem.

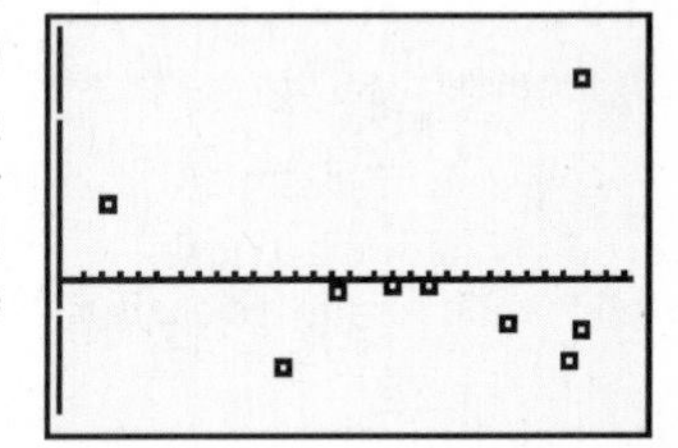

We define a normal plot of the residuals (as in Chapter 4). This normal plot is not a straight line, which indicates a violation of the assumptions. This multiple regression model is not appropriate for these data.

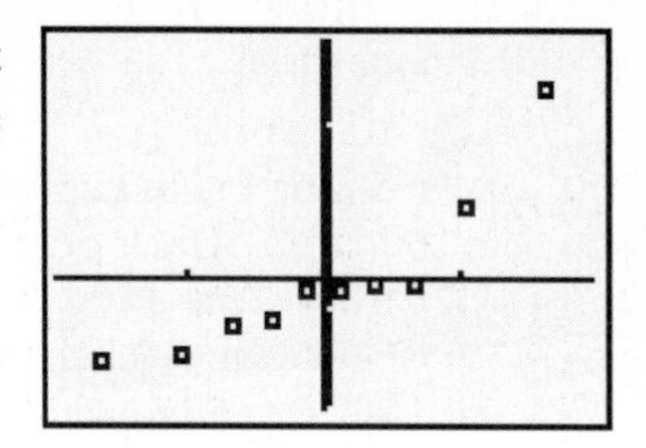

MULTIPLE REGRESSION CONFIDENCE INTERVALS

How well do the midterm grade and number of missed classes predict final grades? Data for a sample of students are below.

Final Grade, y	Midterm Exam, x_1	Classes Missed, x_2
81	74	1
90	80	0
86	91	2
76	80	3
51	62	6
75	90	4
48	60	7
81	82	2
94	88	0
93	96	1

Performing the regression as above, we find the estimated regression equation $Final = 49.41 + 0.502 * Midterm - 4.71 * ClassesMissed$. All coefficients are significantly different from zero and a normal plot of the residuals is relatively straight. We'd like to use the model to create a prediction of the average grade for students with a midterm grade of 75 and 2 absences (a confidence interval); we also want to predict the grade for a particular student with a midterm grade of 75 and 2 absences (a prediction interval).

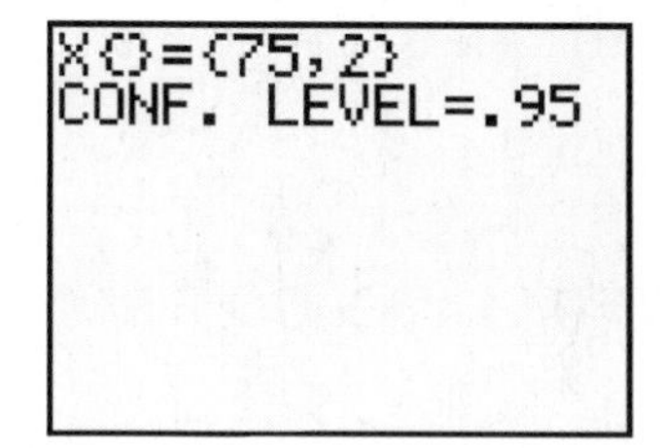

When executing the MULREG program, select 1:YES in answer to the question PREDICT Y. You will be asked to input the independent variables' values. Enter them inside curly braces ([2nd][{] and [2nd][}]) separated by commas. You will then be asked for the desired confidence level.

Pressing [ENTER] performs the calculations and displays the screen at right. We find the point estimate of the final grade for a student with a 75 midterm grade and two absences is 77.66, or 78 (practically speaking). The standard deviation of the fit at that point is 0.392. The confidence interval says we are 95% confident the average final grade for all students with a 75 midterm grade and two absences will be between 76.7 and 78.6. Further, we are 95% confident an individual student with a 75 midterm grade and two absences will earn a final grade between 75.3 and 80. Pressing [ENTER] at this point returns you to the Predict or Quit menu.

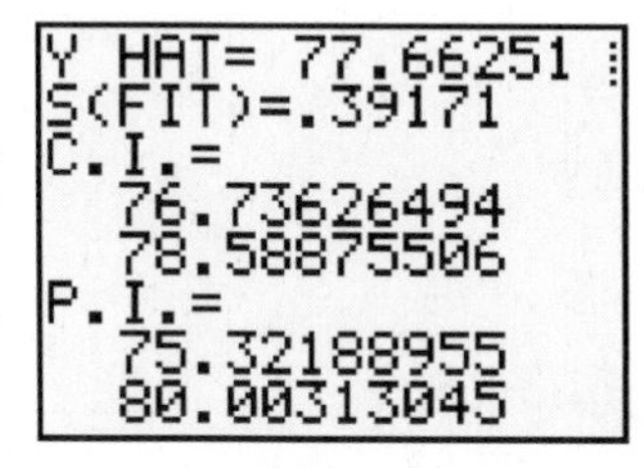

TI-89 Procedure

The data are entered with final grades in `list1`; midterm grades in `list2`, and absences in `list3`. Order of specification in the regression input screen matters. In `list4` we have entered the values for the predictions of interest. They must be entered in the order in which the xlists will be specified.

list1	list2	list3	list4
81	74	1	75
90	80	0	2
86	91	2	------
76	80	3	
51	62	6	
75	90	4	

list4[3]=

Press [2nd][F2] (F7) and select option `8:MultRegInt`. We are first asked the number of independent variables. In our data we have 2. Use the right arrow to access the list of possible values, and select the appropriate one for your data set. Press [ENTER] to continue. Now specify the lists to be used in the regression and the list containing the values to be used in the intervals as at right. Specify the desired confidence level, here 95%, as a decimal.

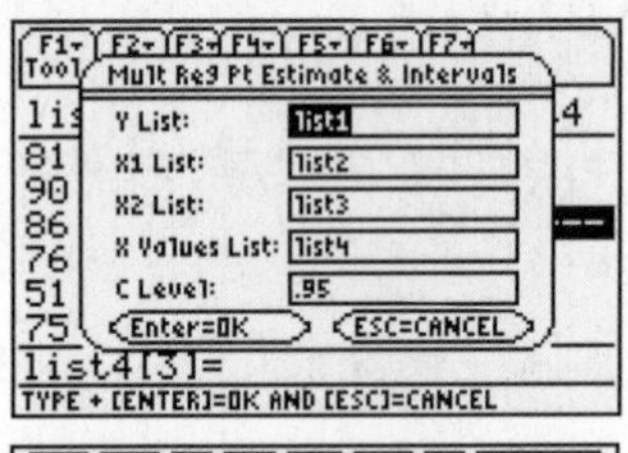

Pressing [ENTER] performs the calculations and displays the screen at right. We find the point estimate of the final grade for a student with a 75 midterm grade and two absences is 77.66 or 78 (practically speaking). The confidence interval says we are 95% confident the average final grade for all students with a 75 midterm grade and two absences will be between 76.7 and 78.6.

Pressing the down arrow several times displays the remainder of the output. We are 95% confident an individual student with a 75 midterm grade and two absences will earn a final grade between 75.3 and 80. The first portion of the coefficients list and the X values used for the intervals are displayed as well.

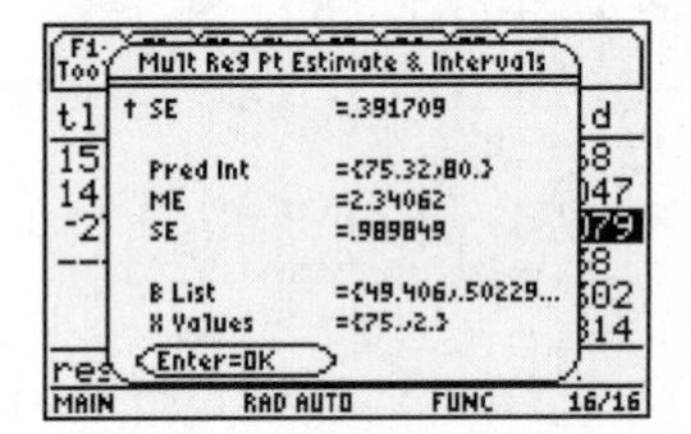

WHAT CAN GO WRONG?

Not much that hasn't already been discussed. The most common errors are misspecification of lists and having more than one plot "turned on" at a time.

How do I get rid of those extra lists?

Press [2nd][+] (MEM). Select `2:Delete`, then select `4:List`. Arrow to the lists to be deleted and press [ENTER].

MULREG Program Listing. (The program can also be downloaded from the text's website.)

```
Disp "DATA IN COLS","OF [A]"
dim([A]) STO▶θ1:Lθ1(1)STO▶N:Lθ1(2) STO▶L:L-1STO▶K
{N,1}STO▶dim([B]):Fill(1,[B])
augment([B],[A]) STO▶ [B]
[B]T[B] STO▶ [D]
seq([D](I,1),I,2,L+1)/NSTO▶Lθ1
{L,L}STO▶dim([C]):AnsSTO▶dim([E])
{N,1}STO▶dim([C])
Input "RESPONSE COL=",R
For(I,1,N):[A](I,R) STO▶ [C](I,1):End
If R≤K:Then
For(J,R+1,L):For(I,1,N)
[B](I,J+1) STO▶ [B](I,J)
End:End
[B]T[B] STO▶ [D]
End
{L,L}STO▶dim([D]):[D]x-1STO▶ [D]
{N,L}STO▶dim([B])
[B]T [C]:[D]AnsSTO▶ [E]
Matr▶list([E],Lθ1)
Disp "COEF Lθ1=":Pause Lθ1
[B][E] STO▶ [B]
Matr▶list([C],LY)
Matr▶list([B],LYP)
mean(LY) STO▶Y
sum((LY-Y)2)STO▶T
sum((LY-LYP) 2STO▶E
LYPSTO▶L5
LY-LYPSTO▶L6
DelVar LY:DelVar LYP
N-LSTO▶M:T-ESTO▶R:R/KSTO▶Q:E/MSTO▶D:√(D) STO▶S:Q/DSTO▶F
S√(seq([D](I,I),I,1,L)) STO▶Lθ2
Disp "STDEV Lθ2=":Pause Lθ2
Lθ1/Lθ2STO▶Lθ3
Disp "T-RATIO Lθ3=":Pause Lθ3
Disp "COEF P Lθ4="
1-2seq(tcdf(0,abs(Lθ3(I)),M),I,1,L) STO▶Lθ4
Pause Lθ4
1-(N-1)*D/TSTO▶A
ClrHome
Output(1,1,"S="
Output(1,9,S
Output(2,1,"R2="
Output(2,9,R/T
Output(3,1,"R2ADJ="
Output(3,9,A
Output(4,1,"REG DF="
Output(4,9,K
```

```
Output(5,1,"ERR DF="
Output(5,9,M
Output(6,1,"TOT DF="
Output(6,9,N-1
Pause
ClrHome
Output(1,1,"SS REG="
Output(1,9,R
Output(2,1,"SS ERR="
Output(2,9,E
Output(3,1,"SS TOT="
Output(3,9,T
Output(4,1,"MS REG="
Output(4,9,Q
Output(5,1,"MS ERR="
Output(5,9,D
Output(6,1,"F=    "
Output(6,9,F
Output(7,1,"P-VAL="
1-Fcdf(0,F,K,M) STO▸P
round(P,5) STO▸P
Output(7,9,P
Pause
ClrHome
Lbl C
Menu("PREDICT Y?","YES",A,"QUIT",B)
Lbl A
Input "X{}=",Lθ5
Input "CONF. LEVEL=",C
{L,1}STO▸dim([C])
1STO▸ [C](1,1)
For(I,1,K):Lθ5(I) STO▸ [C](I+1,1):End
[E]ᵀ[C]
Ans(1,1) STO▸Y
ClrHome
round(Y,5) STO▸Y
Output(1,1,"Y HAT="
Output(1,8,Y
[C]ᵀ[D][C]
Ans(1,1) STO▸V
round(S√(V),5) STO▸T
Output(2,1,"S(FIT)="
Output(2,8,T
TInterval 0,√(M+1),M+1,C
upperSTO▸T
Output(3,1,"C.I.="
Y+TS√(V) STO▸D
Y-TS√(V) STO▸E
Output(4,3,E
Output(5,3,D
Output(6,1,"P.I.="
Y-TS√(1+V) STO▸E
Y+TS√(1+V) STO▸D
```

```
Output(7,3,E
Output(8,3,D
Pause
Goto C
Lbl B
DelVar [C]:DelVar [D]:DelVar [E]:DelVar Lθ5
ClrHome
```